my revision notes

OCR (A) A2 CHEMISTRY

Mike Smith

With thanks to all the students whose valuable feedback helped develop this book.

Hodder Education, an Hachette UK company, 338 Euston Road, London NW1 3BH

Orders

Bookpoint Ltd, 130 Milton Park, Abingdon, Oxfordshire OX14 4SB

tel: 01235 827827

fax: 01235 400401

e-mail: education@bookpoint.co.uk

Lines are open 9.00 a.m.–5.00 p.m., Monday to Saturday, with a 24-hour message answering service. You can also order through the Hodder Education website: www.hoddereducation.co.uk

ISBN 978-1-4441-6299-8

First printed 2012

Impression number 5 4 3

Year 2017 2016 2015 2014

Cover photo reproduced by permission of Frank Peters/Fotolia

Typeset by Dianne Shaw

Printed in India

Hachette UK's policy is to use papers that are natural, renewable and recyclable products and made from wood grown in sustainable forests. The logging and manufacturing processes are expected to conform to the environmental regulations of the country of origin.

P01985

Get the most from this book

Everyone has to decide his or her own revision strategy, but it is essential to review your work, learn it and test your understanding. These Revision Notes will help you to do that in a planned way, topic by topic. Use this book as the cornerstone of your revision and don't hesitate to write in it — personalise your notes and check your progress by ticking off each section as you revise.

☑ Tick to track your progress

Use the revision planner on page 4 to plan your revision, topic by topic. Tick each box when you have:

- revised and understood a topic
- tested yourself
- practised the exam questions and gone online to check your answers and complete the quick quizzes

You can also keep track of your revision by ticking off each topic heading in the book. You may find it helpful to add your own notes as you work through each topic.

My revision planner

Unit F324 Rings, polymers and analysis

	Revised	Tested	Exam ready
1 Rings, acids and amines			
6 Arenes	☐	☐	☐
9 Carbonyl compounds	☐	☐	☐
13 Carboxylic acids and esters	☐	☐	☐
17 Amines	☐	☐	☐

Spectroscopy

NMR spectroscopy — Revised

NMR spectroscopy involves the interaction of nuclei with radio waves, which are at the low-energy end of the electromagnetic spectrum.

If the nucleus of an atom contains an odd number of protons and/or neutrons, the nucleus has a net nuclear spin that can be detected by using radio frequency — for example, ^{1}H and ^{13}C can both be detected.

The nucleus behaves like a tiny bar magnet and as it spins it generates a magnetic moment. Adjacent nuclei also have magnetic moments.

Features to help you succeed

Examiner's tips and summaries

Throughout the book there are tips from the examiner to help you boost your final grade.

Summaries provide advice on how to approach each topic in the exams, and suggest other things you might want to mention to gain those valuable extra marks.

Typical mistake

The examiner identifies the typical mistakes candidates make and explains how you can avoid them.

Now test yourself

These short, knowledge-based questions provide the first step in testing your learning. Answers are at the back of the book.

Definitions and key words

Clear, concise definitions of essential key terms are provided on the page where they appear.

Key words from the specification are highlighted in bold for you throughout the book.

Check your understanding

Use these questions at the end of each section to make sure that you have understood every topic. Answers are at the back of the book.

Exam practice

Practice exam questions are provided for each topic. Use them to consolidate your revision and practise your exam skills.

Online

Go online to check and print your answers to the exam questions and try out the extra quick quizzes at **www.therevisionbutton.co.uk/myrevisionnotes**

My revision planner

Unit F324 Rings, polymers and analysis

Unit F325 Equilibria, energetics and elements

Exam practice answers and quick quizzes at **www.therevisionbutton.co.uk/myrevisionnotes**

Countdown to my exams

6–8 weeks to go

- Start by looking at the specification — make sure you know exactly what material you need to revise and the style of the examination. Use the revision planner on page 4 to familiarise yourself with the topics.
- Organise your notes, making sure you have covered everything on the specification. The revision planner will help you to group your notes into topics.
- Work out a realistic revision plan that will allow you time for relaxation. Set aside days and times for all the subjects that you need to study, and stick to your timetable.
- Set yourself sensible targets. Break your revision down into focused sessions of around 40 minutes, divided by breaks. These Revision Notes organise the basic facts into short, memorable sections to make revising easier.

Revised ☐

4–6 weeks to go

- Read through the relevant sections of this book and refer to the examiner's tips, examiner's summaries, typical mistakes and key terms. Tick off the topics as you feel confident about them. Highlight those topics you find difficult and look at them again in detail.
- Test your understanding of each topic by working through the 'Now test yourself' and 'Check your understanding' questions in the book. Look up the answers at the back of the book.
- Make a note of any problem areas as you revise, and ask your teacher to go over these in class.
- Look at past papers. They are one of the best ways to revise and practise your exam skills. Write or prepare planned answers to the exam practice questions provided in this book. Check your answers online and try out the extra quick quizzes at **www.therevisionbutton.co.uk/myrevisionnotes**
- Try different revision methods. For example, you can make notes using mind maps, spider diagrams or flash cards.
- Track your progress using the revision planner and give yourself a reward when you have achieved your target.

Revised ☐

One week to go

- Try to fit in at least one more timed practice of an entire past paper and seek feedback from your teacher, comparing your work closely with the mark scheme.
- Check the revision planner to make sure you haven't missed out any topics. Brush up on any areas of difficulty by talking them over with a friend or getting help from your teacher.
- Attend any revision classes put on by your teacher. Remember, he or she is an expert at preparing people for examinations.

Revised ☐

The day before the examination

- Flick through these Revision Notes for useful reminders, for example the examiner's tips, examiner's summaries, typical mistakes and key terms.
- Check the time and place of your examination.
- Make sure you have everything you need — extra pens and pencils, tissues, a watch, bottled water, sweets.
- Allow some time to relax and have an early night to ensure you are fresh and alert for the examinations.

Revised ☐

My exams

A2 Chemistry Unit F324

Date: ....

Time: ....

Location: ....

A2 Chemistry Unit F325

Date: ....

Time: ....

Location: ....

1 Rings, acids and amines

Arenes

Structure of benzene

Revised

The French chemist August Kekulé suggested that benzene was a cyclic molecule with alternating C=C double bonds and C–C single bonds.

However, there are three major pieces of evidence against this type of structure:

- Compounds that contain C=C double bonds readily decolorise bromine. Benzene only reacts with bromine when hot and exposed to ultraviolet light or in the presence of a halogen carrier.
- On average, the length of a C–C single bond is 154 pm while the average length of a C=C double bond is 134 pm. All the bonds in benzene are 139 pm. This suggests an intermediate bond somewhere between a double bond and a single bond.
- Experimentally determined enthalpy changes for the hydrogenation of cyclohexene and benzene give a value for benzene that is about 150 kJ mol^{-1} lower than that expected from the alternating double bond-single bond model.

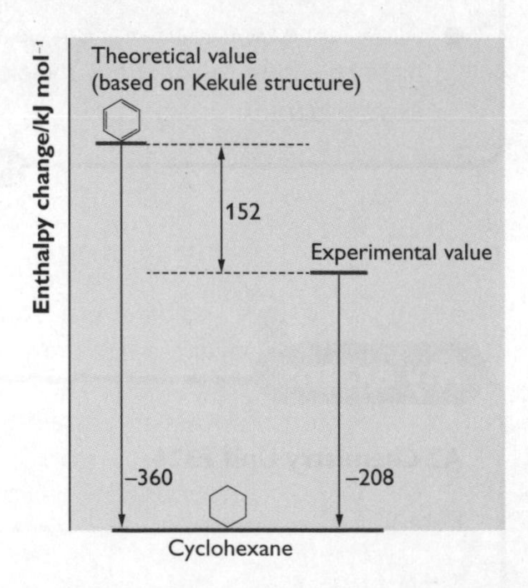

Examiner's tip

All chemistry papers have to examine the seven subsections of *How Science Works*. One of the subsections examines the 'tentative nature of scientific knowledge'. The debate about the Kekulé and delocalised models of benzene structure fits into this category. Questions about this are common.

The current model of benzene is that each carbon atom contributes one electron to a π-delocalised ring of electrons above and below the plane of atoms. Each carbon has one *p*-orbital at right angles to the plane of atoms and adjacent *p*-orbitals overlap so that delocalisation is extended over all six carbon atoms. The π-delocalised ring accounts for the increased stability of benzene as well as explaining the reluctance to react with bromine. In addition, it explains why all six carbon–carbon bond lengths are identical. Benzene is usually represented by the skeletal formula shown below.

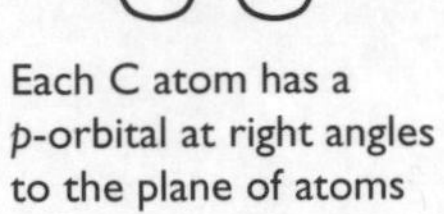

Each C atom has a *p*-orbital at right angles to the plane of atoms

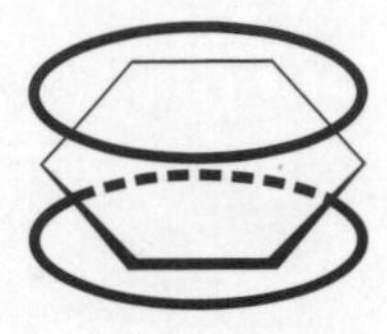

π-delocalised ring above and below the plane

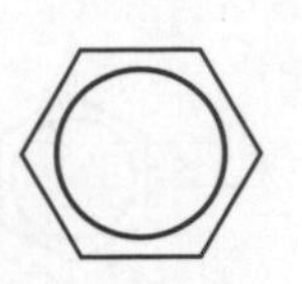

Skeletal formula of benzene

Electrophilic substitution

Revised

The π-delocalised rings make benzene very stable and the electron density in benzene is relatively low. However, benzene does react with **electrophiles** that have a *full* positive charge — an induced dipole in a molecule is not normally sufficient.

An **electrophile** is an electron-pair acceptor that forms a dative covalent bond in a reaction.

Catalysts are used to generate electrophiles such as NO_2^+, Cl^+, Br^+ and CH_3^+. The general equation is:

$$C_6H_6 + X^+ \rightarrow C_6H_5X + H^+$$

where X^+ is the electrophile.

Nitration of benzene

Reagents: HNO_3 and H_2SO_4 (catalyst)
Conditions: approximately 60°C
Balanced equation: $C_6H_6 + HNO_3 \rightarrow C_6H_5NO_2 + H_2O$
Mechanism:

- Generation of the electrophile:

$$H_2SO_4 + HNO_3 \rightleftharpoons HSO_4^- + H_2NO_3^+$$

Sulfuric acid donates a proton to nitric acid

Protonated nitric acid is very unstable and can break down to form

H_2O $^+NO_2$

Nitronium ion

- Electrophilic attack at the benzene ring:

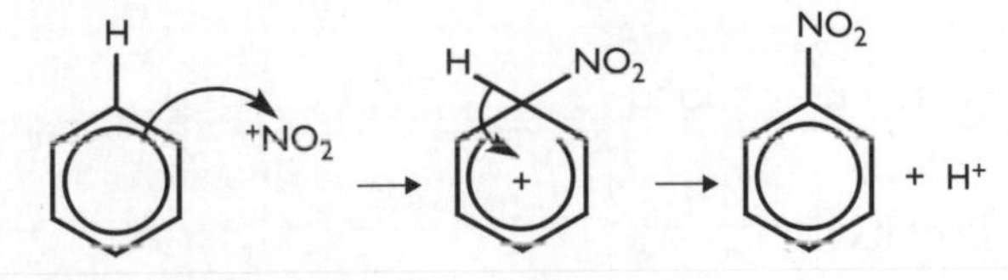

- Regeneration of the catalyst:

$$H^+ + HSO_4^- \rightarrow H_2SO_4$$

Typical mistake

Candidates often take a great deal of care when drawing out the mechanism. They ensure the curly arrows are correctly placed but often lose a mark by forgetting to include the H^+ which is substituted from the ring.

Halogenation of benzene

Reagents: Cl_2 and $AlCl_3$ (catalyst)
Conditions: anhydrous ($AlCl_3$ reacts with water)
Balanced equation: $C_6H_6 + Cl_2 \rightarrow C_6H_5Cl + HCl$
Mechanism:

- Generation of the Cl^+ electrophile:

$$Cl_2 + AlCl_3 \rightarrow Cl^+ + AlCl_4^-$$

- Electrophilic attack at the benzene ring:

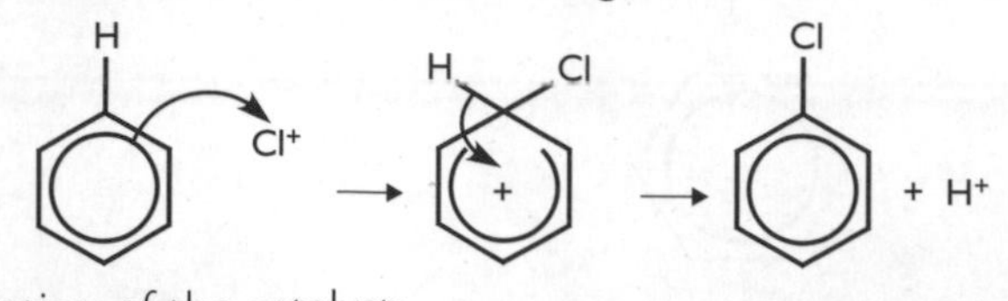

- Regeneration of the catalyst:
  $H^+ + AlCl_4^- \rightarrow AlCl_3 + HCl$

The chlorination of benzene is a Friedel–Crafts reaction in which $AlCl_3$ behaves as a halogen carrier. Halogen carriers are able to accept a halide ion and to 'carry it' through the reaction. At the end of the reaction, the halide ion is released and the hydrogen halide is formed. All aluminium halides, iron(III) halides and iron can behave as halogen carriers.

Bromination of alkenes and arenes

You will recall from Unit F322 that alkenes, such as cyclohexene, react readily with bromine in the absence of sunlight, undergoing **electrophilic addition** reactions.

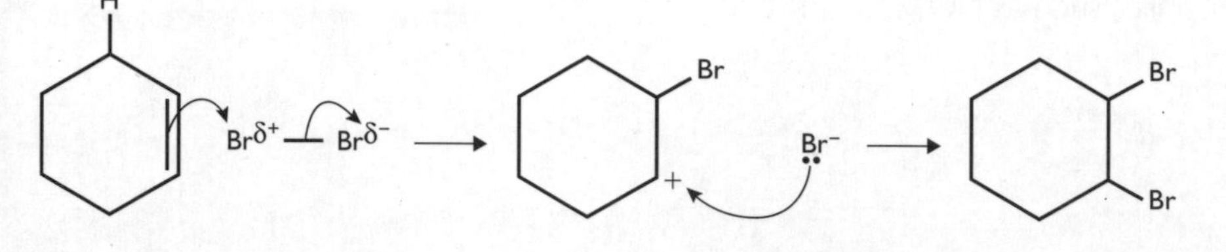

The reaction is rapid and is initiated by the induced dipole in bromine.

Benzene also reacts with bromine but is more resistant, reacting much less readily than an alkene, such as cyclohexene. Benzene requires an electrophile with a full positive charge, Br^+, which is generated in the presence of a halogen carrier. The resultant reaction is electrophilic substitution, *not* electrophilic addition. This is explained by the stability of the π-delocalised ring of electrons that is retained in most reactions of all arenes.

The relative ease of reaction of cyclohexene can be explained by the electron density:

- Cyclohexene has a C=C double bond that has high electron density.
- The π-electrons are localised between the two carbon atoms in the C=C double bond.
- The electron density is sufficient to induce a dipole in the Br–Br bond and an electrophile is generated.
- The electrophile is attracted to cyclohexene because of its high electron density.

Uses of arenes

Revised

Arenes such as benzene, methylbenzene and 1,4-dimethylbenzene are used as additives to improve the performance of petrol. They are manufactured by reforming straight-chain alkanes.

Hexane

C_6H_{14} ⟶ (benzene) + $4H_2$

Benzene is the feedstock for a variety of products ranging from medicines, such as aspirin and benzocaine, to explosives such as 2,4,6-trinitromethylbenzene (TNT) and a range of azo dyes. Phenylethene (styrene) is also manufactured from benzene and is the monomer used to produce the polymer poly(phenylethene) or polystyrene.

Products made from benzene are of great value. However, benzene itself is carcinogenic and may cause leukaemia. Chlorinated benzene compounds are extremely toxic.

Now test yourself

1 Methylbenzene, $C_6H_5CH_3$, can be nitrated to form $CH_3C_6H_4NO_2$. Draw and name the isomers.

2 Methylbenzene, $C_6H_5CH_3$, can be further nitrated to form 2,4,6-trinitromethylbenzene. Write a balanced equation for this reaction.

Answers on p. 92

Phenols

In phenols, the –OH group is attached directly to the benzene ring. Phenol behaves as a weak **acid**. It also undergoes electrophilic substitution reactions much more readily than benzene.

Phenol forms **salts** by reaction with both sodium and sodium hydroxide.

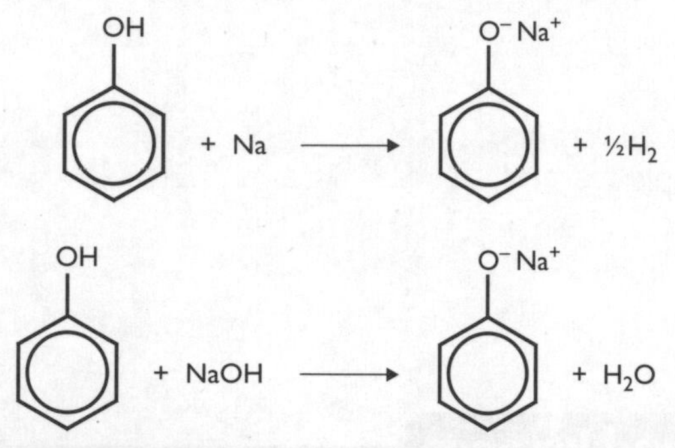

Now test yourself

3 What is the molecular formula of 4-methylphenol?

Answers on p. 92

An **acid** is a proton donor.

A **salt** is formed when an acid has one or more of its hydrogen ions replaced by a metal ion or an ammonium ion, NH_4^+.

Examiner's tip

If you remember the definition of a salt it is easy to work out the formula of an organic salt. Simply replace the acidic H^+ in the phenol (or carboxylic acid) with the metal ion (usually Na^+).

Phenol reacts readily with bromine. The bromine is decolorised and white crystals of 2,4,6-tribromophenol are formed.

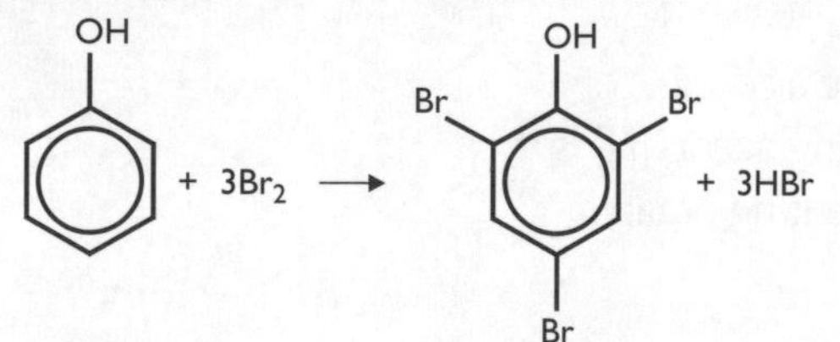

Unlike benzene, phenol does not require a halogen carrier and reacts instantly with bromine. This is explained by the activation of the ring due to delocalisation of one of the lone pairs of electrons on the oxygen atom into the ring. This increases the electron density which in turn polarises the halogen and increases the attraction for the halogen. The result is increased reactivity of phenol.

Uses of phenols

Phenols are used in the production of disinfectants and antiseptics such as 2,4,6-trichlorophenol (TCP) and in the production of plastics and resins.

Carbonyl compounds

Formation of aldehydes and ketones

The presence of the carbonyl group, C=O, in a molecule means that it is unsaturated. The position of the C=O group on the carbon chain determines whether or not the molecule is classified as an aldehyde or a ketone. Aldehydes *always* have the C=O at the end of the carbon chain.

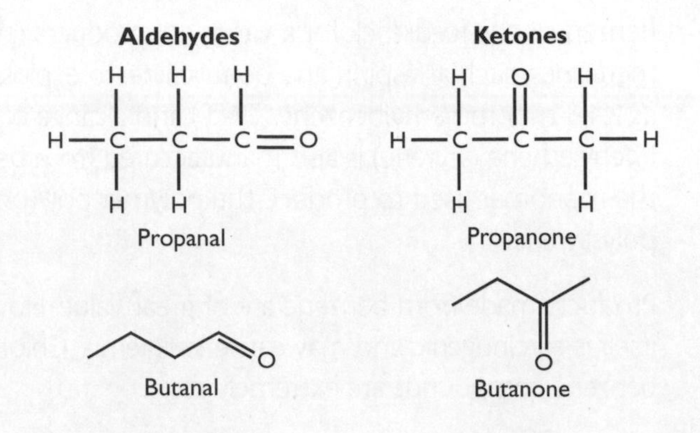

You met the carbonyl group in Unit F322 when you studied the chemistry of alcohols. Aldehydes and ketones are formed by the oxidation of alcohols using $Cr_2O_7^{2-}/H^+$ (e.g. $K_2Cr_2O_7/H_2SO_4$). This is bright orange and changes to green during the redox process. When oxidising a primary alcohol the choice of apparatus is important. Refluxing produces a carboxylic acid; distillation produces an aldehyde.

Oxidation of a primary alcohol to an **aldehyde** by distillation, for example:

$$CH_3CH_2OH + [O] \rightarrow CH_3CHO + H_2O$$

Ethanol → Ethanal

Oxidation of a primary alcohol to a **carboxylic acid** by refluxing, for example:

$$CH_3CH_2OH + 2[O] \rightarrow CH_3CO_2H + H_2O$$

Ethanol → Ethanoic acid

Oxidation of a secondary alcohol to a **ketone,** for example:

$$CH_3CHOHCH_3 + [O] \rightarrow CH_3COCH_3 + H_2O$$

Propan-2-ol → Propan-2-one

Reactions common to both aldehydes and ketones

Revised

Reduction

Aldehydes and ketones can be reduced to their respective alcohols. Sodium tetrahydridoborate(III), $NaBH_4$, is a suitable reducing agent. [H] is used to represent the reducing agent in equations representing organic reduction reactions.

Aldehydes are reduced to primary alcohols, for example:

$$CH_3CH_2CHO + 2[H] \rightarrow CH_3CH_2CH_2OH$$

Ketones are reduced to secondary alcohols, for example:

$$CH_3COCH_3 + 2[H] \rightarrow CH_3CH(OH)CH_3$$

$NaBH_4$ is the source of the hydride ion, $H:^-$, which is a **nucleophile**. The intermediate formed reacts with the solvent, H_2O, to form the alcohol.

A **nucleophile** is an electron-pair donor that forms a dative covalent bond in a reaction.

The mechanism of the reduction of ethanal with aqueous $NaBH_4$ is shown below:

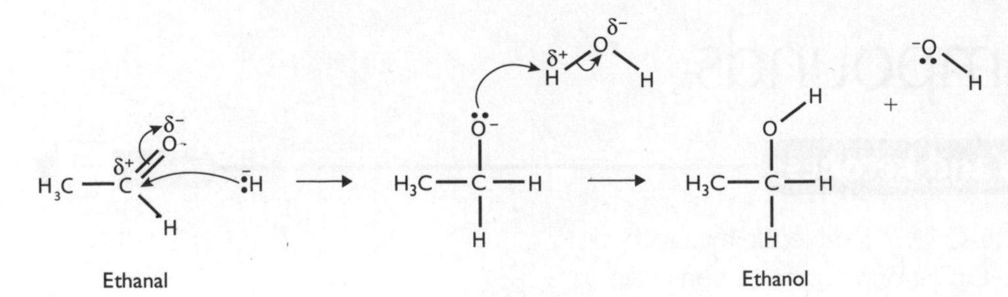

Ketones behave similarly. The mechanism of the reduction of propanone with aqueous $NaBH_4$ is shown below:

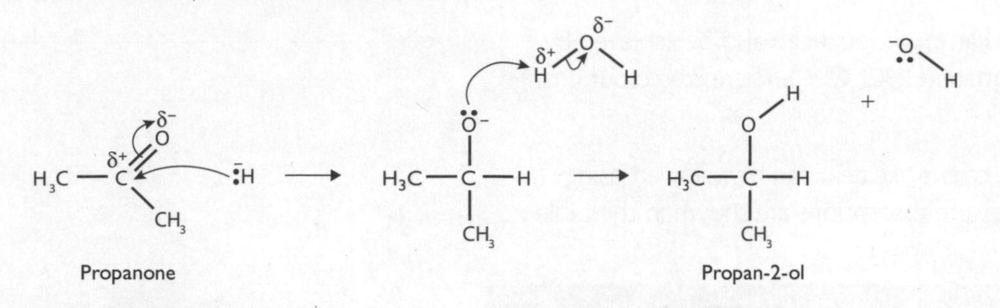

Examiner's tip

The nucleophilic addition mechanism is simplified if an acid, $H^+(aq)$, is added. The mechanism becomes:

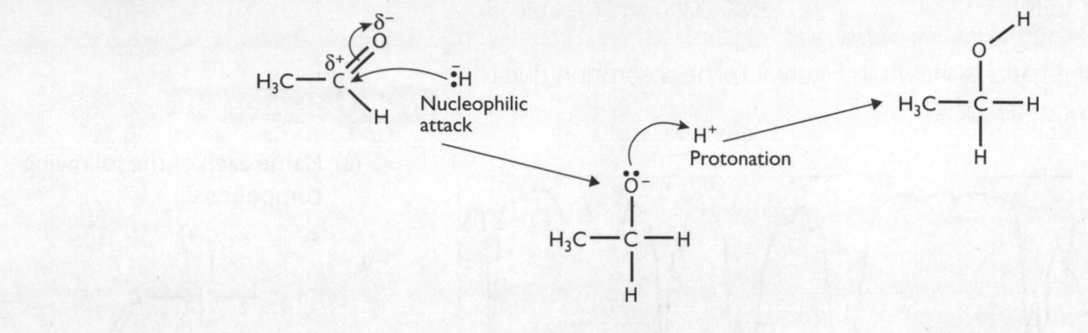

Characteristic tests for carbonyl compounds

Revised

Reaction with 2,4-dinitrophenylhydrazine

Candidates are not expected to be able to recall the formula of 2,4-dinitrophenylhydrazine (the abbreviation 2,4-DNPH is acceptable). The reactions with 2,4-DNPH) are important for two reasons:

- 2,4-DNPH reacts with a carbonyl compound to produce a distinctively coloured precipitate, which is usually bright red, orange or yellow. Therefore, this reaction can be used to identify the presence of a carbonyl group.
- The brightly coloured organic product (the 2,4-DNPH derivative) is relatively easy to purify by recrystallisation. Therefore, the melting point of the brightly coloured precipitate can be measured. Each derivative has a different melting point, the value of which can be used to identify the specific carbonyl compound.

Reactions of aldehydes alone

Revised

Aldehydes and ketones can be distinguished by a series of redox reactions. Aldehydes are readily oxidised to carboxylic acids whereas ketones are not easily oxidised.

Aldehydes react with Tollens' reagent, which is an aqueous solution of Ag^+ ions in an excess of ammonia, $Ag(NH_3)_2^+$. When Tollens' reagent is reacted with an aldehyde and warmed gently in a water bath at about 60°C, silver metal is precipitated which forms a distinctive silver mirror. This is a redox reaction — Ag^+ ions are reduced to silver metal and the aldehyde is oxidised to a carboxylic acid.

The Ag^+ ion gains an electron and is, therefore, reduced to silver

The aldehyde is oxidised to a carboxylic acid

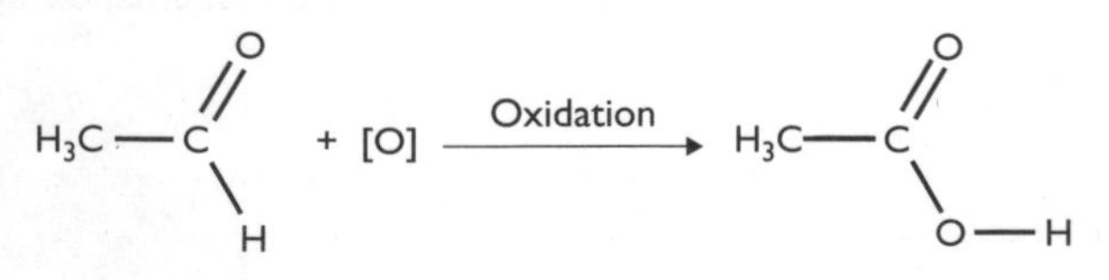

Tollens' reagent does not react with ketones because they are not readily oxidised.

Oxidation of an aldehyde to a carboxylic acid can also be achieved by refluxing with acidified dichromate ($H^+/Cr_2O_7^{2-}$). There is a colour change from orange to green.

Oxidation of an aldehyde to a carboxylic acid can be followed using infrared spectroscopy. The relevant absorptions are shown in the table below.

Group	Compounds	IR absorption
C=O	Aldehydes, ketones, carboxylic acids	1680–1750 cm^{-1}
O–H	Carboxylic acids	2500–3300 cm^{-1} (very broad)

The IR spectrum of ethanal is shown in Figure 1.1. The absorption due to the carbonyl group can be seen.

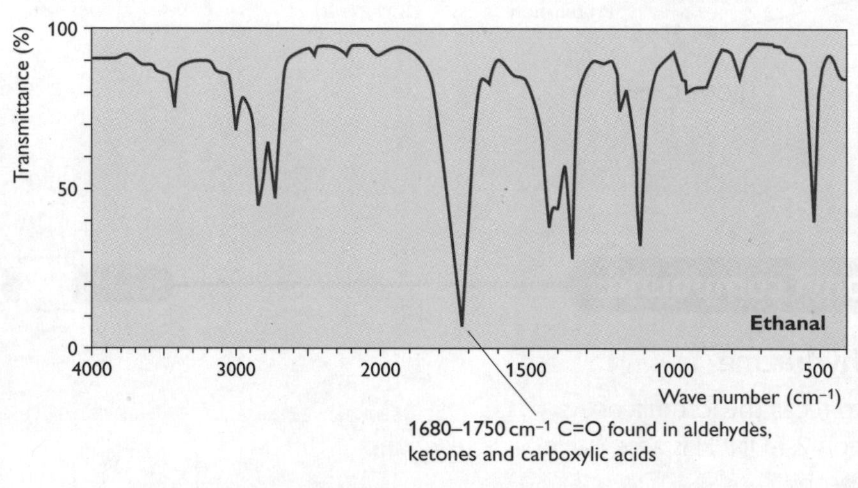

Figure 1.1 Infrared spectrum of ethanal

The IR spectrum of ethanoic acid is shown in Figure 1.2. This spectrum shows the absorptions for both the C=O and O–H groups.

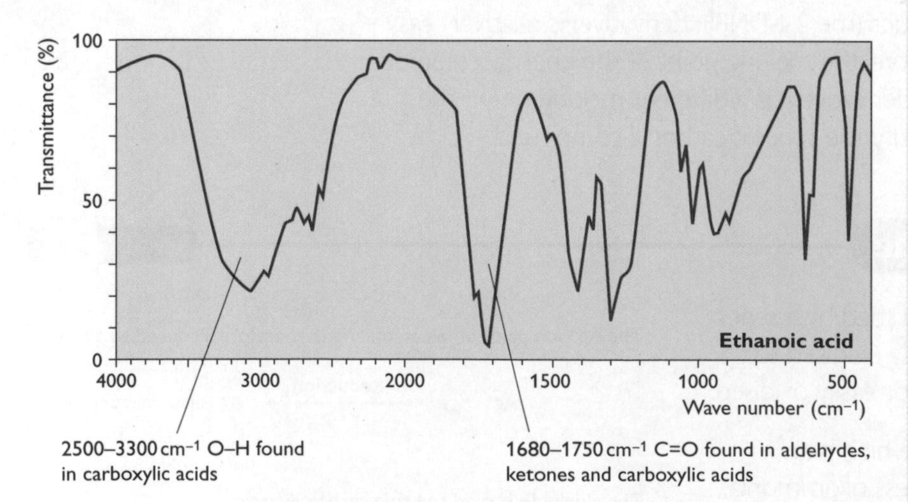

Figure 1.2 Infrared spectrum of ethanoic acid

Now test yourself

4 (a) Name each of the following compounds:

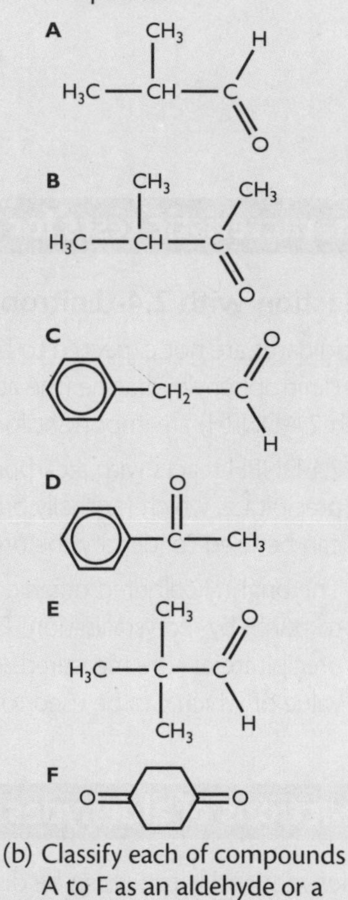

(b) Classify each of compounds A to F as an aldehyde or a ketone.

(c) What is the molecular formula of compound F?

(d) Identify the alcohols from which C and D could be prepared.

(e) Compounds B and E are isomers of each other. Draw three other isomers of B and E.

Answers on p. 92

Carboxylic acids and esters

Carboxylic acids

Revised

All carboxylic acids contain the functional group:

The carboxylic acid group can be attached to either a chain (aliphatic) or to a ring (aromatic), for example:

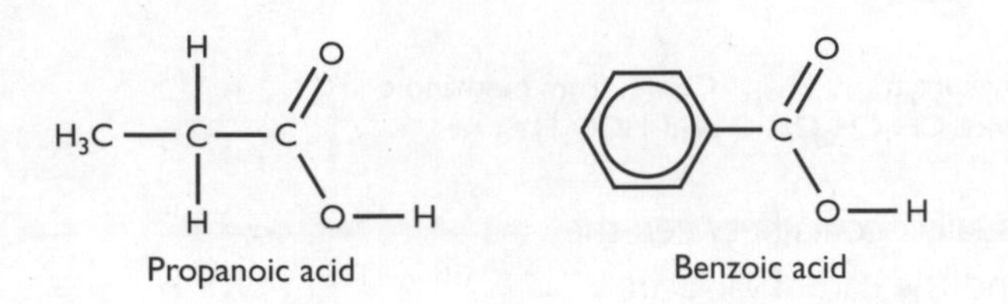

Carboxylic acids such as methanoic acid and ethanoic acid are soluble in water. The solubility is explained by the ability of carboxylic acids to form hydrogen bonds with water.

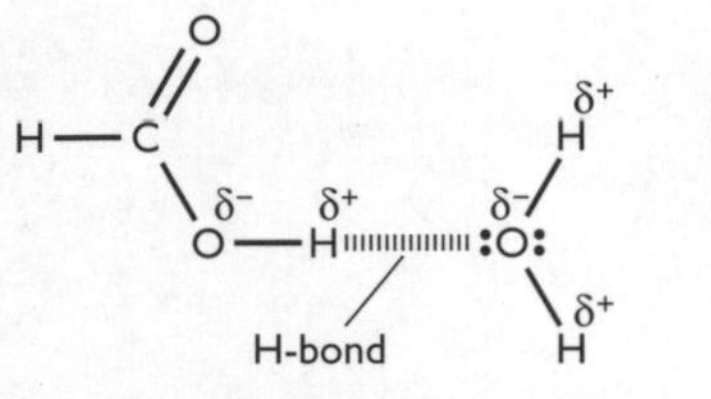

Methanoic acid is soluble in water because it can form H-bonds with water molecules

The solubility of carboxylic acids decreases with increased molar mass.

Carboxylic acids are acidic. Therefore they donate protons. However, they are weak acids, so dissociate only partially into their ions:

$$CH_3CO_2H(aq) \rightleftharpoons CH_3CO_2^-(aq) + H^+(aq)$$

They show typical reactions of acids and can form salts (carboxylates). In each of the examples that follows, the acid is ethanoic acid and the salt formed is sodium ethanoate:

- acid + base → salt + water
  $CH_3CO_2H(aq) + NaOH(aq) \rightarrow CH_3CO_2^-Na^+(aq) + H_2O(l)$
- acid + (reactive) metal → salt + water
  $CH_3CO_2H(aq) + Na(s) \rightarrow CH_3CO_2^-Na^+(aq) + \frac{1}{2}H_2(g)$
- acid + carbonate → salt + water + carbon dioxide
  $CH_3CO_2H(aq) + Na_2CO_3(aq) \rightarrow CH_3CO_2^-Na^+(aq) + H_2O(l) + CO_2(g)$

The reaction with a carbonate is used as a test for a carboxylic acid. When an acid is added to a solution of a carbonate, bubbles (effervescence/fizzing) of carbon dioxide are seen.

Carboxylic acids react with alcohols to form esters. This type of reaction is known as **esterification**. It is reversible and is usually carried out in the presence of an acid catalyst, such as concentrated sulfuric acid. The general reaction can be summarised as follows:

Now test yourself

5 Write a balanced equation, including state symbols, for the reaction of HCO_2H with:

(a) Na(s)

(b) $NaHCO_3(aq)$

Answers on p. 92

Examiner's tip

The reaction of a carboxylic acid with sodium produces bubbles, but it cannot be used as a test for a carboxylic acid because phenols also react with sodium to produce a salt and bubbles (in this case, of hydrogen).

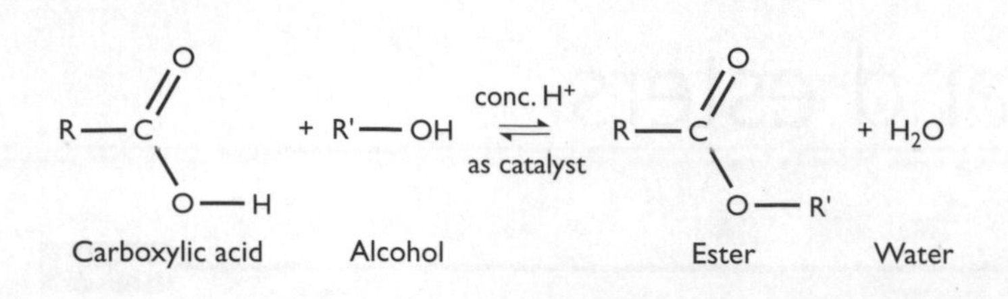

> **Typical mistake**
>
> Candidates often quote H_2SO_4(aq) as the catalyst and lose the mark because (aq) indicates that water is present. Water is a product in the esterification reaction. The presence of additional water would drive the equilibrium to the left and inhibit the formation of the ester.

Esters are named from the alcohol and the carboxylic acid from which they are derived. The first part of the name relates to the alcohol and the second part of the name relates to the acid, for example:

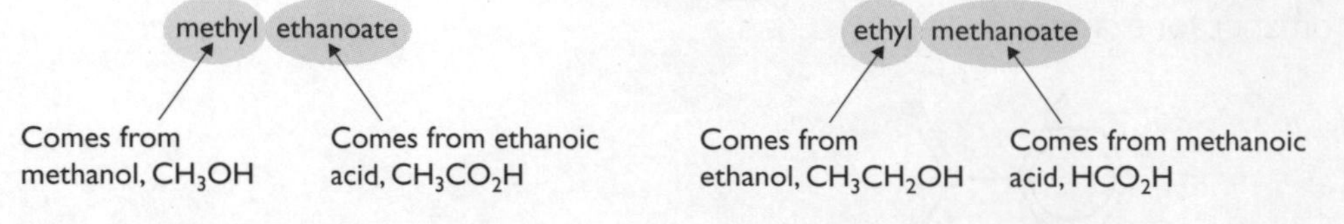

When organic compounds react, the reaction usually occurs between the two functional groups, in this case the alcohol and the carboxylic acid. It is helpful to draw the two reacting molecules with the functional groups facing each other:

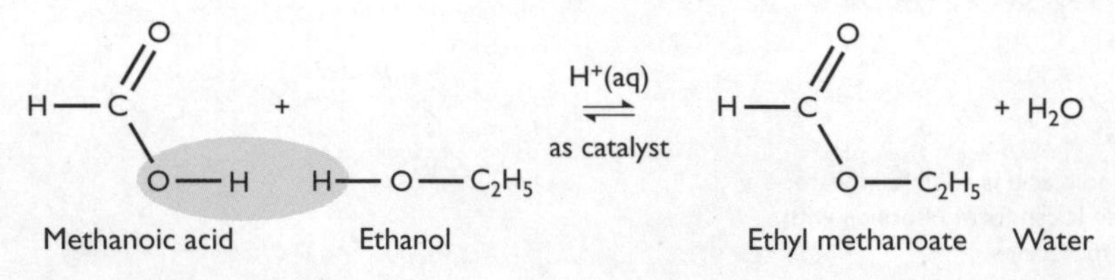

Esters are used in flavourings and perfumes. They often contribute to the flavour and aroma associated with fruits.

Esters

Revised

Esters react with water. The **hydrolysis** reaction is slow and is carried out under reflux in the presence of either an acid, H^+(aq), or a **base**, OH^-(aq). Acid-catalysed hydrolysis leads to the formation of the carboxylic acid and the alcohol:

> A **base** is a proton acceptor.

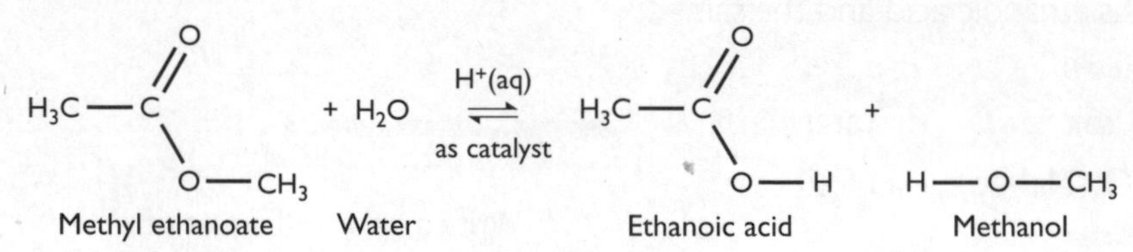

Base-catalysed hydrolysis leads to the formation of the salt of the carboxylic acid (the carboxylate) and the alcohol:

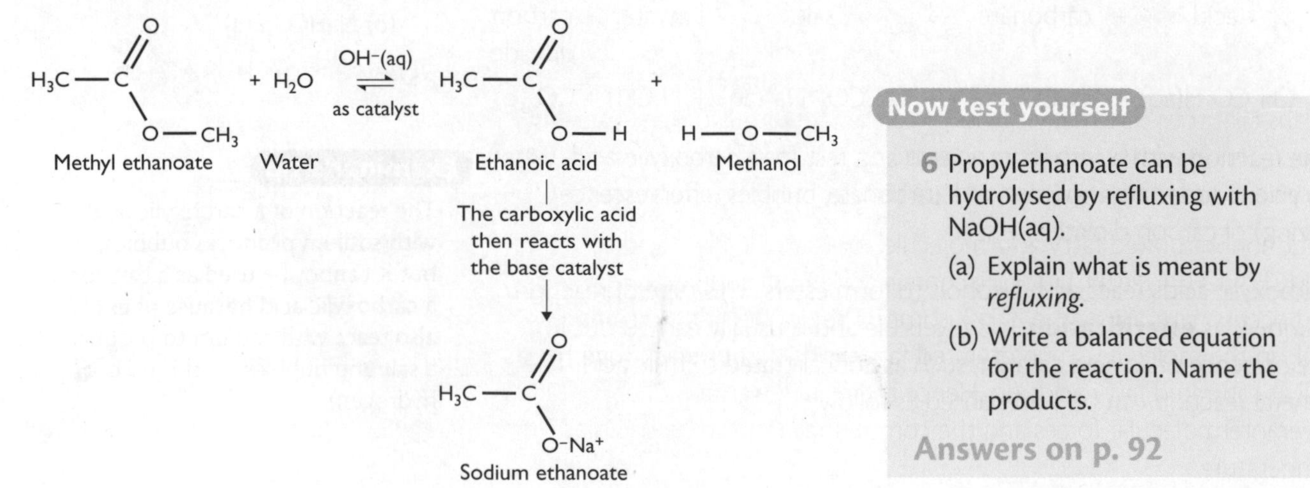

> **Now test yourself**
>
> **6** Propylethanoate can be hydrolysed by refluxing with NaOH(aq).
>
> (a) Explain what is meant by *refluxing*.
>
> (b) Write a balanced equation for the reaction. Name the products.
>
> Answers on p. 92

Many esters occur naturally in plants and animals. The esters found in vegetable oils (derived from plants) and in animal fats are esters of fatty acids.

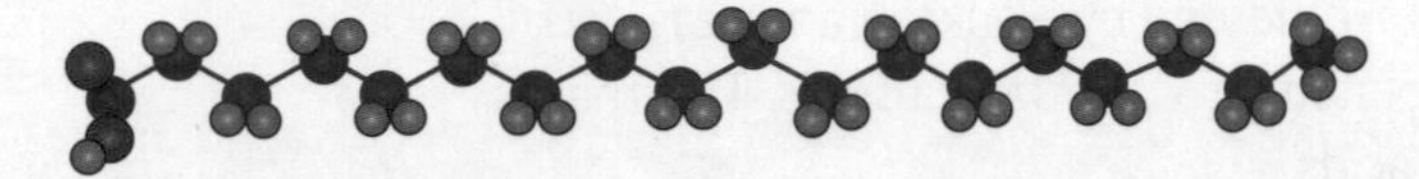

Fatty acids have trivial names that are in common use — for example oleic acid, which is found in olive oil. However, they also have systematic names. The systematic name of oleic acid is *Z*-octadec-9-enoic acid, 18, 1(9) which sounds much more difficult than oleic acid. However, the numbers at the end of the name indicate that a molecule of oleic acid contains 18 carbon atoms and that there is one *Z* (*cis*) double bond on the ninth carbon starting from the carboxylic acid end. Therefore, the formula is $CH_3(CH_2)_7CH{=}CH(CH_2)_7COOH$.

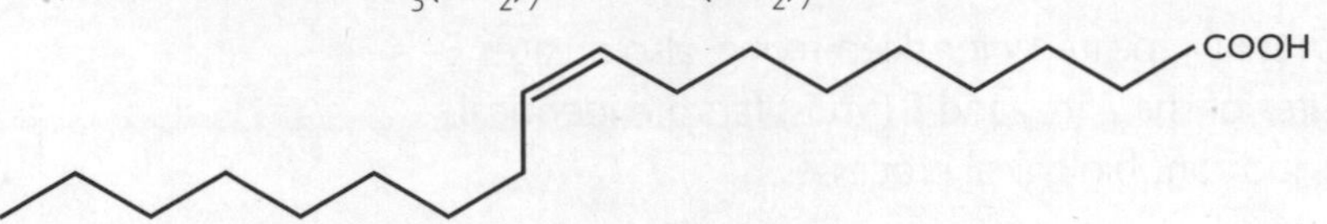

Unsaturated and saturated fats

Revised

Fatty acids are an essential part of a healthy diet. They can be saturated, monounsaturated or polyunsaturated. A saturated fatty acid molecule contains no C=C double bonds and is essentially a straight chain.

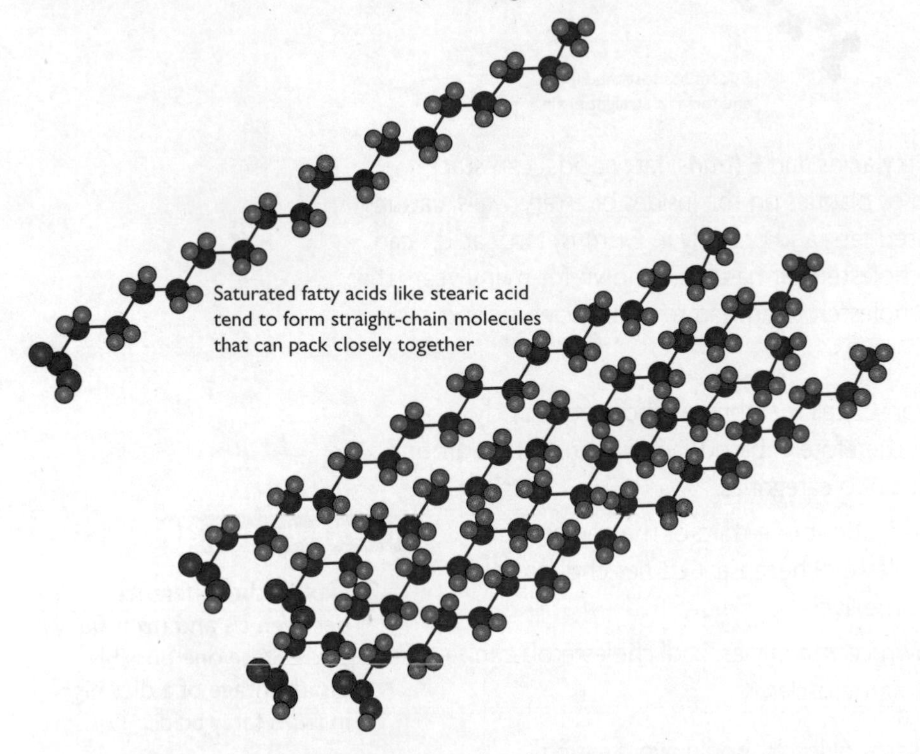

Saturated fatty acids like stearic acid tend to form straight-chain molecules that can pack closely together

Saturated fatty acid molecules are straight chains that pack together closely. This means that there are many points of contact, resulting in significant intermolecular forces. This increases viscosity and reduces volatility. Therefore, saturated fatty acids are relatively dense and are solid at room temperature.

A monounsaturated fatty acid molecule, for example oleic acid, $CH_3(CH_2)_7CH{=}CH(CH_2)_7COOH$, contains one C=C double bond. Oleic acid occurs naturally as the *Z* (*cis*) isomer. The double bond creates a 'kink' in the molecule, so unsaturated fatty acids do not stack together as well as saturated fatty acids do. Hence there are fewer points of contact, fewer intermolecular forces and the compounds tend to be liquid at room temperature.

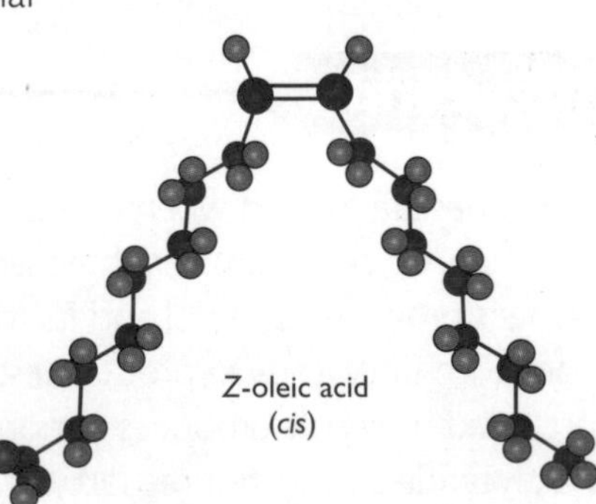

Z-oleic acid (*cis*)

A polyunsaturated fatty acid molecule contains more than one C=C double bond, so there is more than one 'kink 'in the molecule. Therefore polyunsaturated fatty acids stay fluid even at low temperature. They are synthesised by plankton and seaweed, which are eaten by fish and become incorporated into their fatty tissues. This makes fish a ready source of polyunsaturated fatty acids.

Naturally occurring fatty acids, found in both animals and plants, are mostly the *Z* (*cis*) form and therefore have the distinctive kinked shape. Food manufacturers partially hydrogenate (add hydrogen to) some of the C=C double bonds because this prolongs the shelf life of the product — the partially hydrogenated fatty acids are less likely to become rancid. The hydrogenation process removes some of the kinks, so the long chain is straightened out.

Exposure to prolonged heat, such as using oil repeatedly for deep-frying, also creates *E* (*trans*) fatty acids. The chemical structures of the *Z* (*cis*) and *E* (*trans*) fats are identical, but their geometric differences affect important biological processes.

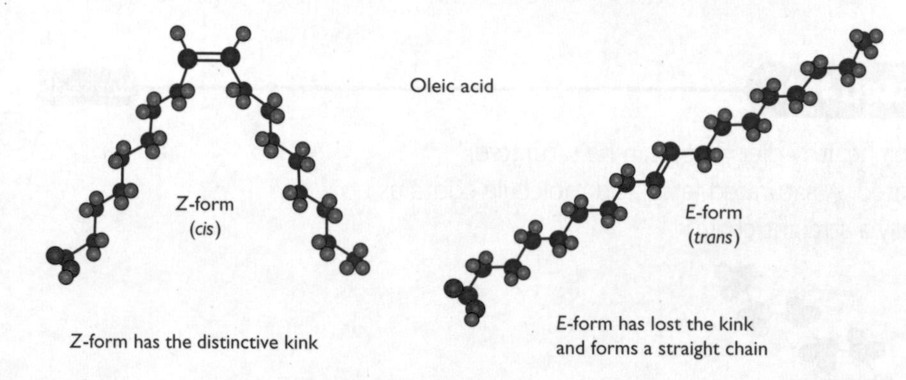

Straight-chain fatty acids (saturated fatty acids and *E* (*trans*) fatty acids) can stack together, which can lead to a build-up of plaques on the insides of artery walls, causing them to thicken. A diet high in saturated fats, and possibly in *E* (*trans*) fatty acids, can also lead to increased blood levels of cholesterol. It has been known for many years that such a diet, high in saturated fat and cholesterol, can lead to atherosclerosis and heart disease.

Cholesterol is essential for the formation of cell membranes and several body hormones. It is insoluble in water (and therefore in blood) and is transported around the body by lipoproteins. These fall into two categories:

- High-density lipoprotein, HDL, carries about one-third of the cholesterol. It is known commonly as 'good' cholesterol because it carries cholesterol away from the arteries and back to the liver.
- Low-density lipoprotein, LDL, (known commonly as 'bad' cholesterol) can lead to a build-up of plaques and to atherosclerosis.

People worried about their blood cholesterol levels should avoid eating saturated fatty acids and *E* (*trans*) fatty acids by using liquid vegetable oils.

Now test yourself

7 Explain the difference between *cis* and *trans* fatty acids. State one possible disadvantage of a diet high in *trans* fatty acids.

Answers on p. 93

Triglycerides

Revised

Glycerides are formed when one or more fatty acid molecules react with a glycerol molecule. Glycerol contains three alcohol groups and is known as a triol; its systematic name is propane-1,2,3-triol and its formula is $CH_2(OH)CH(OH)CH_2OH$. Glycerol can react with fatty acids to produce esters. Reaction of a molecule of glycerol with one fatty acid molecule produces a monoglyceride, with two fatty acid molecules produces a diglyceride and with three fatty acid molecules produces a triglyceride:

Propane-1,2,3-triol (glycerol) backbone

Ester link

Naturally occurring fats and oils (triglycerides) can be hydrolysed by refluxing with a base. This produces propane-1,2,3-triol and the salts of the fatty acids. This is known as **saponification**, which means 'the forming of soap'. Modern soaps are made from blends of oils. The base hydrolysis of a triglyceride is shown below:

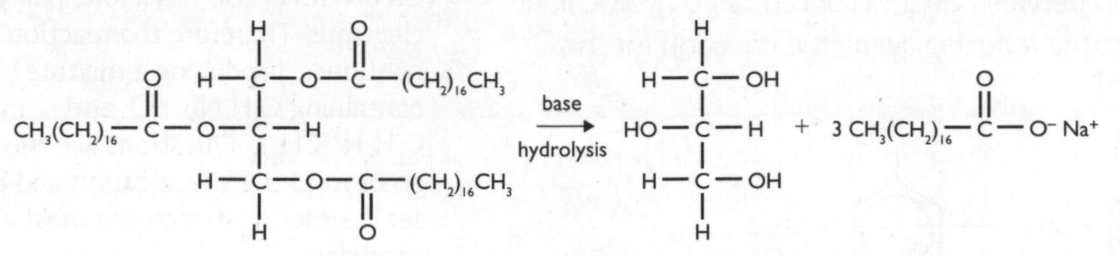

Vegetable oils and animal fats and their derivatives, particularly methyl esters, are being used increasingly as alternative diesel fuels, known as **biodiesel**. Biodiesel is defined as 'a monoalkyl ester of long-chain fatty acids from renewable feedstocks such as vegetable oil or animal fats'.

Typical mistake

When candidates are asked to draw a triglyceride they often draw the ester group incorrectly (left). The correct orientation of the ester group is shown on the right.

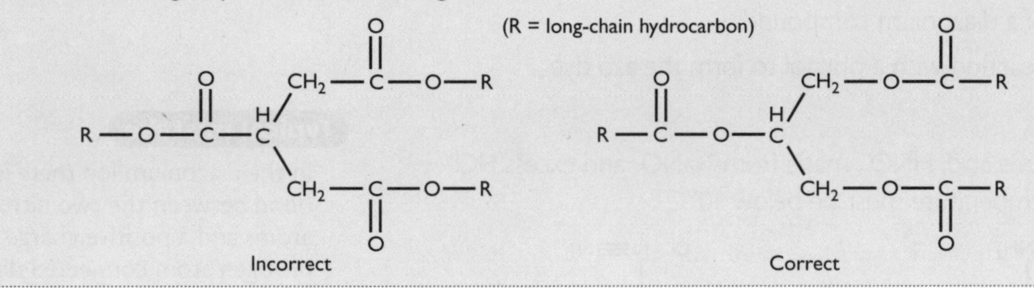

Amines

Basicity of amines

Revised

Amines are weak bases since the lone pair of electrons on the nitrogen can accept a proton from water to form a dative covalent bond:

$$C_2H_5NH_2 + H_2O \rightarrow (C_2H_5NH_2{:}\rightarrow H)^+ + OH^-$$

Consequently, amines react with acids to form salts. Two examples are:

$C_2H_5NH_2 + HCl \rightarrow C_2H_5NH_3^+Cl^-$
Ethylamine

$C_6H_5NH_2 + H_2SO_4 \rightarrow C_6H_5NH_3^+HSO_4^-$
Phenylamine

In the second reaction, $(C_6H_5NH_3^+)_2SO_4^{2-}$ could also be formed:

$$2C_6H_5NH_2 + H_2SO_4 \rightarrow (C_6H_5NH_3^+)_2SO_4^{2-}$$

Preparation of amines

Revised

Primary aliphatic amines such as ethylamine, $CH_3CH_2NH_2$, can be prepared by the reaction between a halogenoalkane and excess ethanolic ammonia:

$$CH_3CH_2Cl + NH_3\ (alc) \rightarrow CH_3CH_2NH_2 + HCl$$

The reaction mixture has to be heated under pressure.

Aromatic amines such as phenylamine, $C_6H_5NH_2$, can be prepared by heating nitrobenzene, under reflux, with tin and concentrated hydrochloric acid. This is a reduction reaction in which the reducing agent is formed from the reaction between tin and concentrated hydrochloric acid. Using [H] to represent the reducing agent, the equation for this reaction is:

NO_2 (on benzene ring) + 6[H] → NH_2 (on benzene ring) + $2H_2O$

Examiner's tip

An amine has a lone pair of electrons on the nitrogen and can therefore behave as a nucleophile. For example, $C_2H_5NH_2$ reacts with CH_3Cl to produce $C_2H_5NHCH_3$ and HCl. The nitrogen in the product, $C_2H_5NHCH_3$, also has a lone pair of electrons. Therefore the reaction continues, producing a mixture containing $C_2H_5N(CH_3)_2$ and $C_2H_5N^+(CH_3)_3$. Questions like this go beyond the specification and are set in order to stretch the most able candidates.

Synthesis of azo dyes

Revised

Aromatic amines, such as phenylamine, are essential for the industrial synthesis of azo dyes. The synthesis involves two stages:

- formation of a diazonium compound
- a coupling reaction with a phenol to form the azo dye

Stage 1
Reagents: nitrous acid, HNO_2, made from $NaNO_2$ and excess HCl
Conditions: temperature must be below 10°C
Equation:

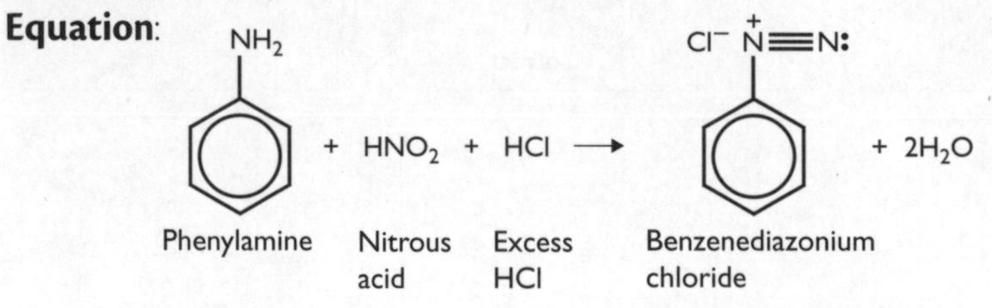

The reaction has to kept below 10° C because benzenediazonium chloride is unstable and reacts readily with water to produce phenol, N_2 and HCl.

Typical mistake

In the diazonium ion there is a triple bond between the two nitrogen atoms and a positive charge on the nitrogen atom connected directly to the ring. Many candidates put the positive charge on the wrong nitrogen; others 'float' it between the two nitrogens, hoping that the examiner will be generous. You will only get the mark if the charge is located correctly.

Stage 2
Reagents: phenol
Conditions: alkaline solution (in the presence of $OH^-(aq)$)
Equation:

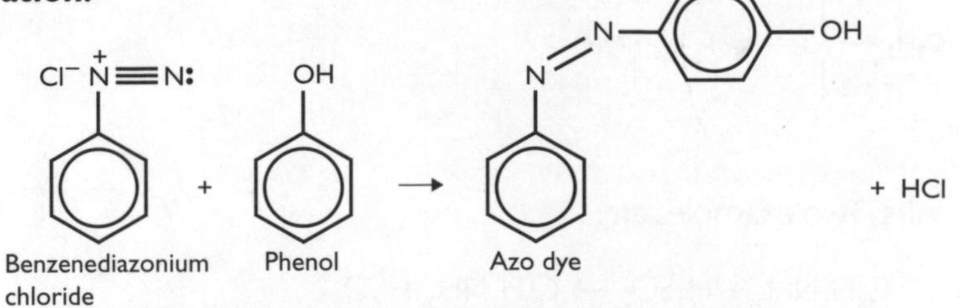

The –N=N– group absorbs light, making azo compounds brightly coloured. The exact colour depends on the other substituents on the aromatic ring.

Now test yourself

8 Diazonium compounds can be formed from aromatic amines. Azo dyes are formed by the coupling reaction between a diazonium compound and a phenol. Copy and complete the table below by drawing the diazonium compound and the azo dyes.

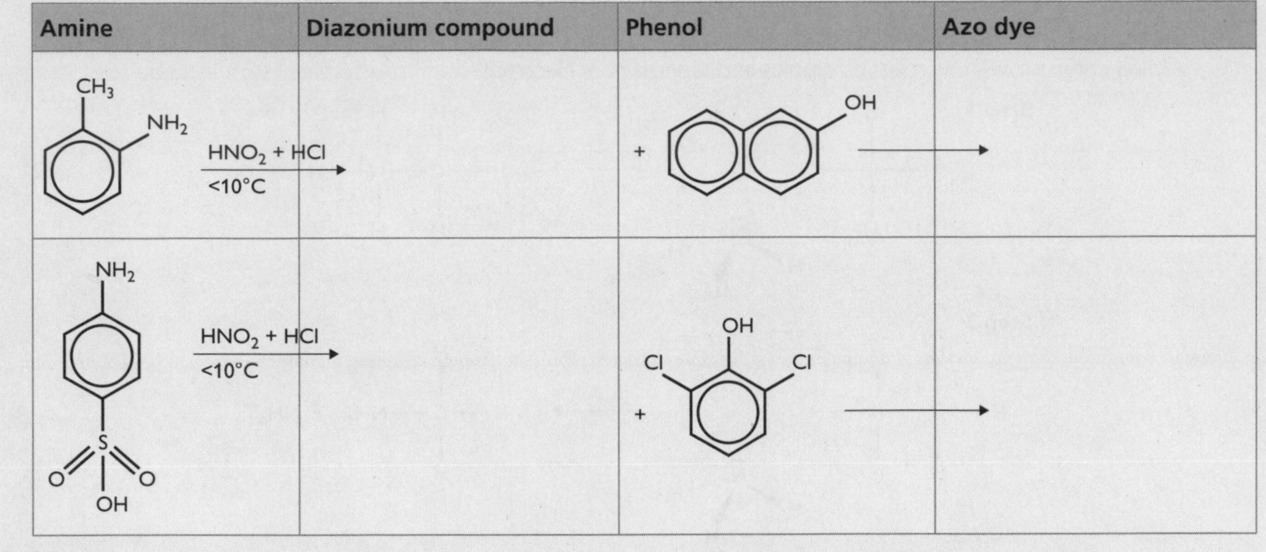

Amine	Diazonium compound	Phenol	Azo dye
CH_3, NH_2 — HNO_2 + HCl, <10°C →		+ OH →	
NH_2, S, O, O, OH — HNO_2 + HCl, <10°C →		+ OH, Cl, Cl →	

Answers on p. 93

Check your understanding

1 Benzene, phenol and cyclohexene all react with bromine.

(a) For each reaction:
- name the type of reaction
- state the reagents and conditions (if any)
- identify the organic product and state any observations

(b) Explain the different rates of reaction of bromine with each of the following pairs of compounds:
  (i) benzene and phenol
  (ii) benzene and cyclohexene

2 Butan-2-ol can be prepared by the reduction of a carbonyl compound. Identify the carbonyl compound, and state the reagents and conditions. Write a balanced equation and, with the aid of curly arrows, describe the mechanism.

3 12.0 g of ethanoic acid were reacted with methanol in the presence of concentrated sulfuric acid. 3.70 g of the ester, methylethanoate, were isolated.

(a) Write a balanced equation for the reaction and explain the role of the concentrated sulfuric acid.
(b) Calculate the percentage yield.
(c) Calculate the atom economy.
(d) Suggest why the percentage yield is so low.

4 Benzene reacts with alkenes in the presence of an acid catalyst in the following way:

$$C_6H_6 + C_nH_{2n} \xrightarrow{H^+ \text{ catalyst}} C_6H_5C_nH_{2n+1}$$

The initial step in the reaction is the H^+ catalyst reacting with the alkene to form a carbonium ion, ${}^+C_nH_{2n+1}$.

In the presence of an acid catalyst, benzene reacts with propene to form an aromatic hydrocarbon with molecular formula C_9H_{12}.

(a) Outline, with the aid of curly arrows, the reaction between benzene and propene in the presence of H^+.
(b) Explain why it is possible to form two isomers of C_9H_{12}.
(c) Alkyl groups such as CH_3 are electron releasing. They can push electrons along a σ-bond. This is called an inductive effect and can stabilise a carbonium ion. Use this and your understanding of the mechanism to explain why the major product of the reaction between benzene and propene is $C_6H_5CH(CH_3)_2$ and not $C_6H_5CH_2CH_2CH_3$.

5 Aliphatic amines can be prepared by the reaction between chloroalkanes and ammonia. The ammonia has to be dissolved in ethanol.

(a) Explain why aqueous ammonia cannot be used in this reaction.

(b) Suggest why ammonia dissolved in a saturated aqueous solution of NH_4Cl might be suitable.

(c) Chloromethane reacts with NH_3 to produce methylamine. Copy and complete the following mechanism by adding curly arrows, any relevant dipoles and lone pairs of electrons.

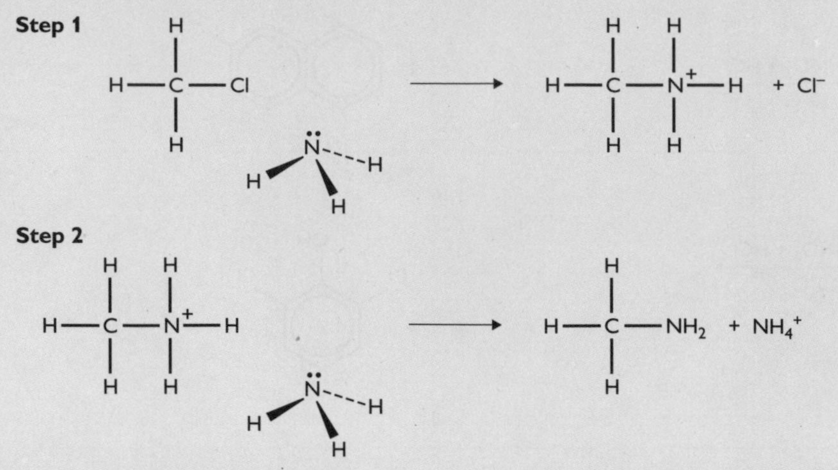

(d) In the reaction between chloroethane and ammonia a variety of different organic products is formed, including $CH_3CH_2NH_2$, $(CH_3CH_2)_2NH$, $(CH_3CH_2)_3N$ and $(CH_3CH_2)_4N^+$. Suggest how you might manipulate the reaction conditions to ensure that you obtain mainly:

(i) $CH_3CH_2NH_2$

(ii) $(CH_3CH_2)_4N^+$

6 Diazonium compounds are useful synthetic chemicals that can be used to prepare a variety of other compounds, such as azo dyes. However during their preparation the temperature has to be controlled carefully. If the diazonium compound is subsequently warmed in the presence of water a phenol is formed together with nitrogen and hydrogen chloride. Use this information and your knowledge of the reactions in the specification to suggest how benzene might be converted to 4-methylphenol. Identify any likely impurities or by-products.

7 A carbonyl compound that contains at least one hydrogen atom on the carbon atom adjacent to the carbonyl group, (i.e. CHC=O) can, in the presence of a base such as NaOH, undergo a condensation reaction. For example, propanone forms 4-hydroxy-4-methylpentan-2-one:

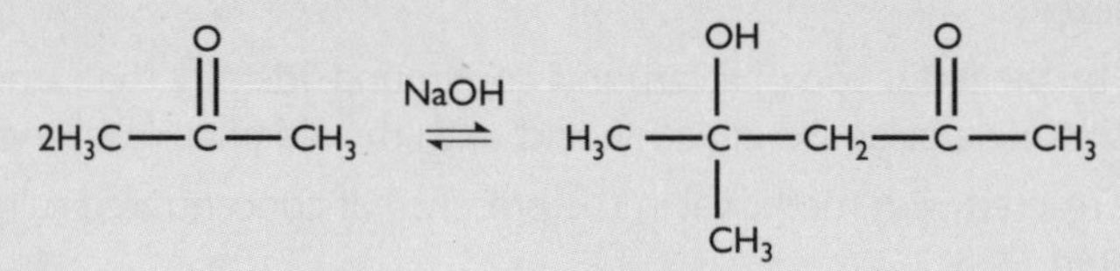

Use this information and your knowledge of the reactions in the specification to suggest how but-2-enal, $CH_3CHCHCHO$, could be made from ethanal.

Answers on pp. 93–95

Exam practice

1 Methylbenzene is an important industrial chemical. It is used in the production of polyurethane plastic foams and fibres such as Lycra®. The production of such foams and fibres involves the nitration of methylbenzene.

(a) Methylbenzene undergoes electrophilic substitution with the nitronium ion, NO_2^+, to form 4-nitromethylbenzene, $CH_3C_6H_4NO_2$.

(i) With the aid of curly arrows, show the mechanism for the formation of 4-nitromethylbenzene. [3]

(ii) In an experiment, 9.20 g of methylbenzene were used and 5.48 g of pure 4-nitromethylbenzene were isolated. Calculate the percentage yield and the atom economy of the reaction. [5]

(b)

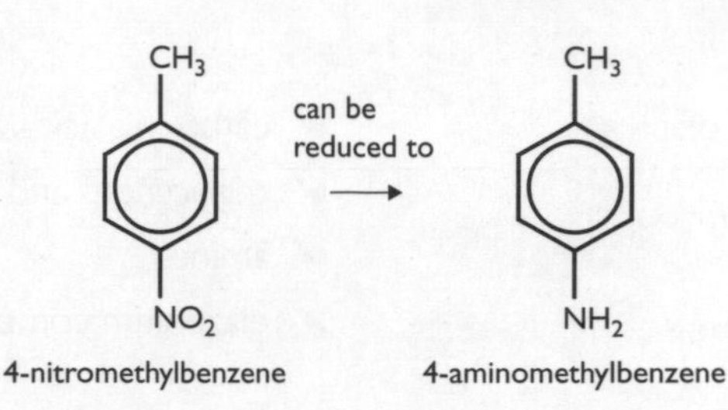

(i) Suggest a suitable reducing agent or a suitable reducing mixture for this reaction. [1]

(ii) Construct a balanced equation for this reduction. Use [H] to represent the reducing agent. [2]

(c) There are six structural isomers of dinitromethylbenzene, $CH_3C_6H_3(NO_2)_2$. Four are drawn for you, draw the structure of the other two isomers. [2]

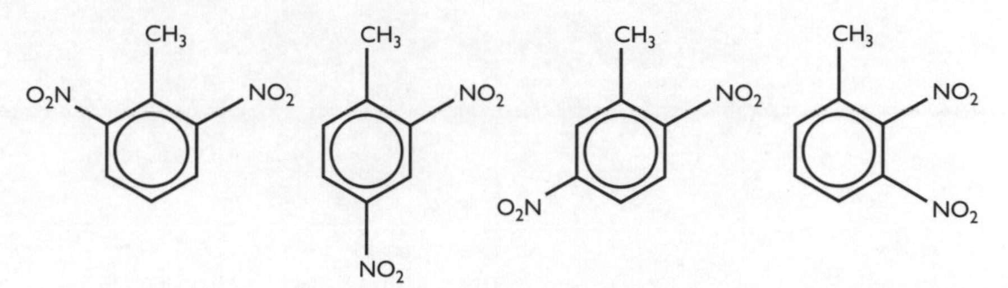

(d) The manufacture of Lycra® involves one of these six isomers. A small section of Lycra® is shown below.

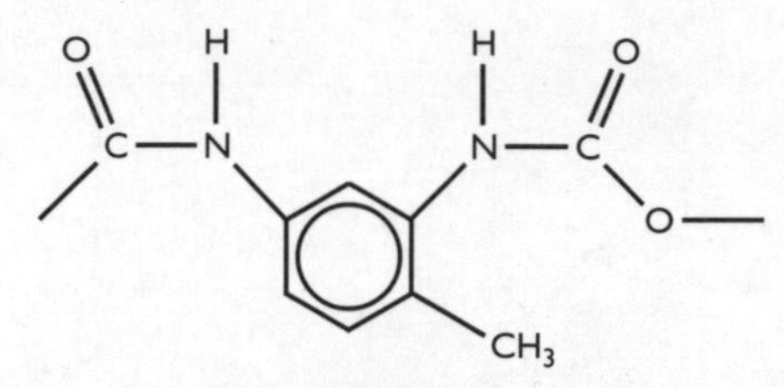

Draw the structure of the isomer of dinitromethylbenzene used in the manufacture of Lycra®. [1]

2 Propan-1-ol can be oxidised to both propanal and to propanoic acid.

(a) (i) State a suitable oxidising mixture. [2]

(ii) State what you would see during the oxidations. [1]

Using [O] to represent the oxidising mixture, write balanced equations to show the oxidation of:

(iii) propan-1-ol to propanal [1]

(iv) propan-1-ol to propanoic acid [1]

(v) Describe a simple *chemical* test to distinguish between propanal and propanoic acid. State what you would see. [2]

(b) Compound X contains carbon, hydrogen and oxygen only. The relative molecular mass of compound X is 102.0. When 5.1 g of compound X is burnt in excess oxygen, 6.0 dm^3 of CO_2 is produced. Compound X can be hydrolysed to form propan-1-ol and one other organic compound. Use *all* of the information in the question to deduce the molecular formula of compound X. Draw the structure of compound X. Show all your working. [5]

3 Compound A has the structure shown below.

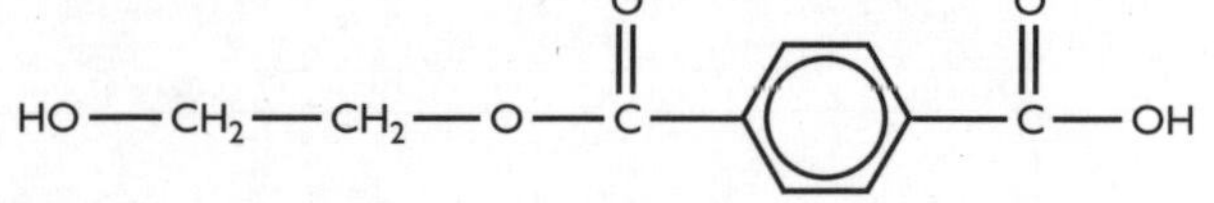

(a) Deduce the empirical formula and molecular formula of compound A. [2]

(b) Suggest three different reagents that will react with compound A. Identify the organic product(s) of each reaction and name the type of reaction involved. [9]

4 (a) Nitrobenzene can be converted into benzenediazonium chloride, $C_6H_5N_2Cl$. For each step, state the reagents and conditions, and write an equation. Show the structure of the organic products. [6]

(b) Benzenediazonium chloride reacts with a chlorinated phenol to form an azo dye with a relative molecular mass of 267.0 and the following composition by mass: C, 53.9%; H, 3.0%; N,10.5%; Cl, 26.6%; O, 6.0%. Use this information to deduce the structure of the azo dye. [4]

Answers and quick quiz 1 online

Online

Examiner's summary

You should now have an understanding of:

- ✔ structure of benzene
- ✔ electrophilic substitution
- ✔ reactions of phenol
- ✔ reactions of carbonyl compounds
- ✔ carboxylic acids and esters
- ✔ triglycerides and fatty acids
- ✔ amines
- ✔ diazonium compounds and azo dyes

2 Polymers and synthesis

Amino acids and chirality

Amino acids

Revised

The general formula of an α-**amino acid** is $RCH(NH_2)COOH$, where R represents the side-chain.

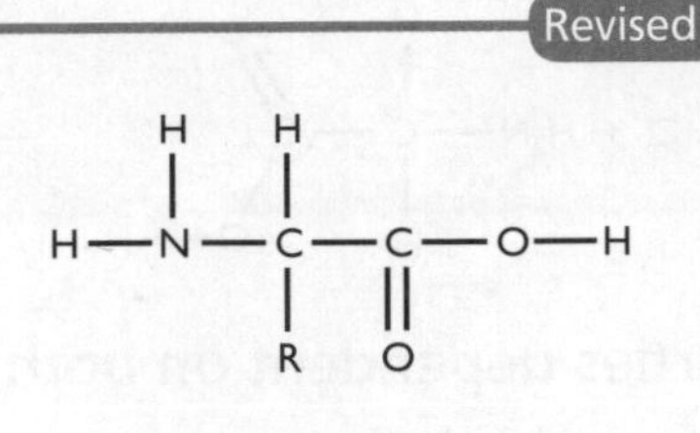

The simplest α-amino acid is aminoethanoic acid, or glycine, where the R group is hydrogen. In 2-aminopropanoic acid, or alanine, the R group is CH_3. Alanine has two optical isomers; glycine is not optically active. This is because alanine has an asymmetric (**chiral**) carbon atom whereas glycine does not.

The general formula of an α-**amino acid** is $RCH(NH_2)COOH$. The amine group (NH_2) and the carboxylic acid group (COOH) are bonded to the same carbon atom.

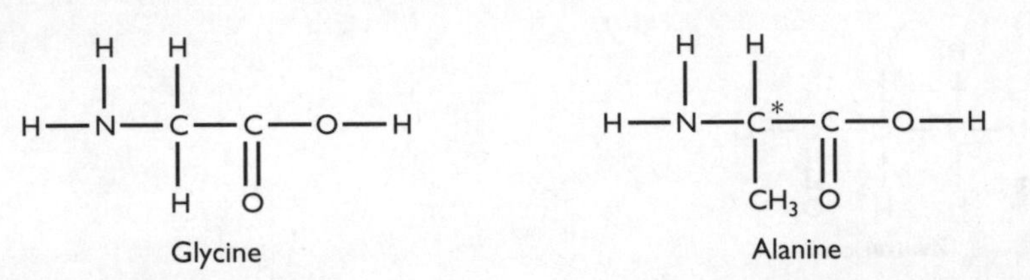

A **chiral** (asymmetric) carbon atom is a carbon atom that is bonded to four different atoms or groups. Glycine is not optically active because the carbon atom is bonded to two hydrogen atoms. The optical isomers of alanine are shown opposite.

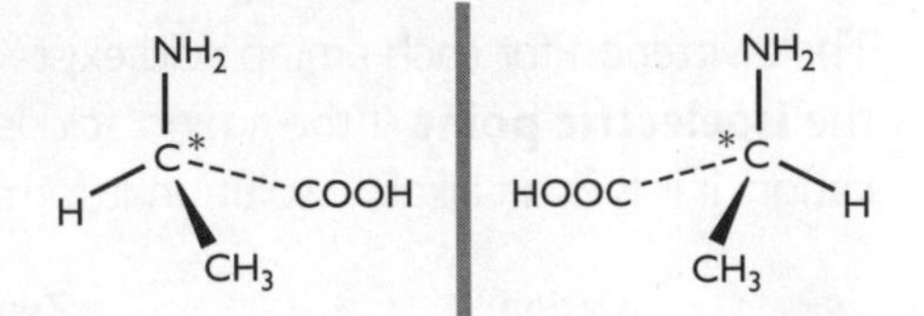

Typical mistake

Candidates lose marks for careless drawings of isomers. The diagrams below both lose marks. Compare them with the correct diagrams above.

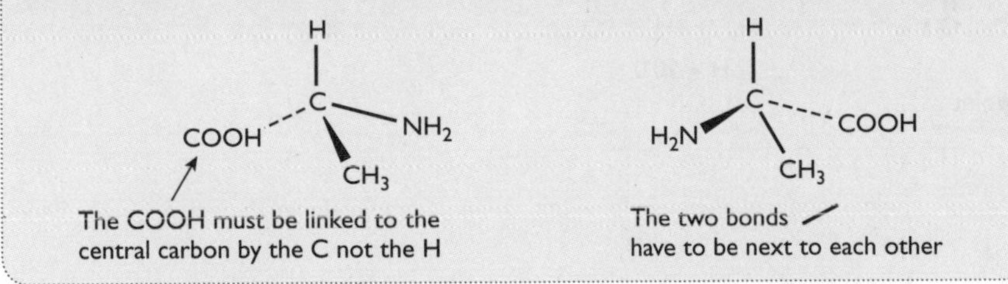

Some amino acids have two chiral carbon atoms and therefore have four (2^2) optical isomers.

Amino acids are bi-functional because they contain two functional groups — carboxylic acid and amine.

Behaving as a carboxylic acid

An amino acid can react with a base to produce a salt, for example:

$$CH_2NH_2CO_2H(aq) + NaOH(aq) \rightarrow CH_2NH_2CO_2^-Na^+(aq) + H_2O(l)$$
Glycine

An amino acid can react with an alcohol to produce an ester, for example:

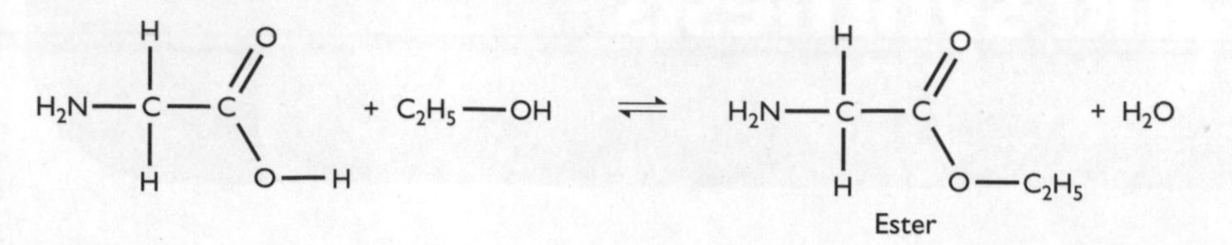

Behaving as an amine

An amino acid can also behave as a primary amine and will react with an acid to produce a salt

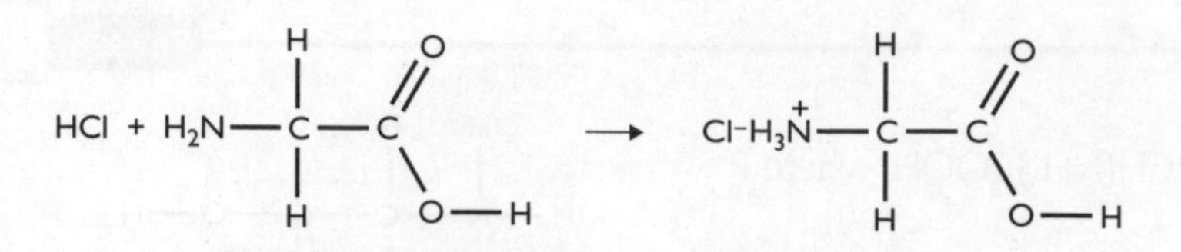

Properties dependent on both functional groups

Amino acids also display properties that depend on both functional groups. Unlike most organic compounds amino acids tend to have high melting points and are water soluble. This is due to the formation of **zwitterions**.

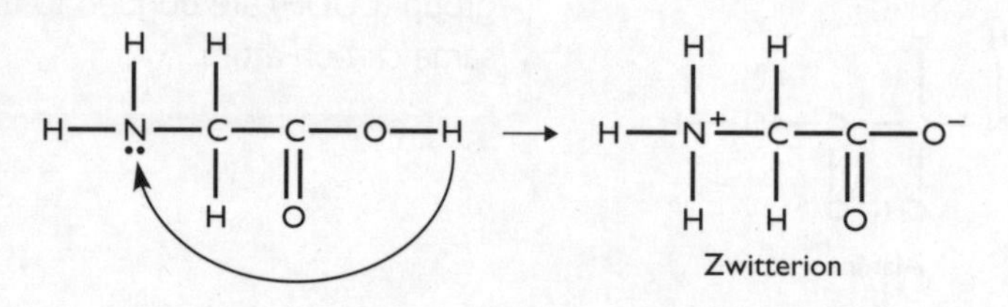

The zwitterion for each amino acid exists at a particular pH, known as the **isoelectric point**. If the amino acid is in an acidic solution it forms a cation. If it is in an alkaline solution it forms an anion.

The **isoelectric point** is the pH at which the amino acid forms the zwitterion.

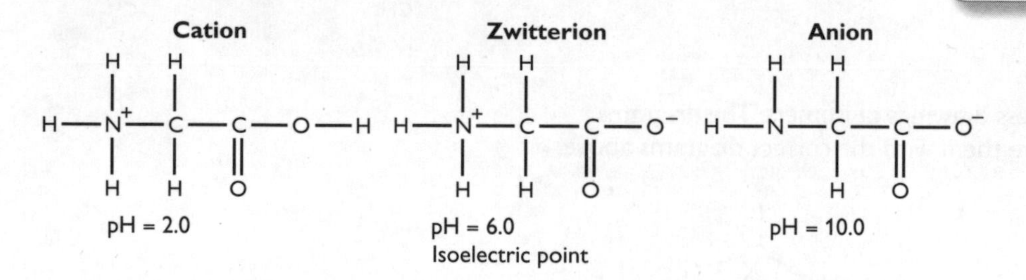

Amino acids can react to form peptides:

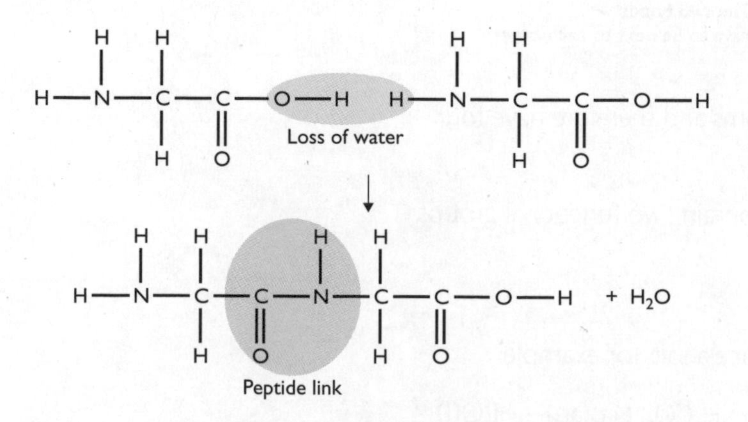

If two different amino acids such as glycine (gly) and alanine (ala) react it is possible to form two different dipeptides:

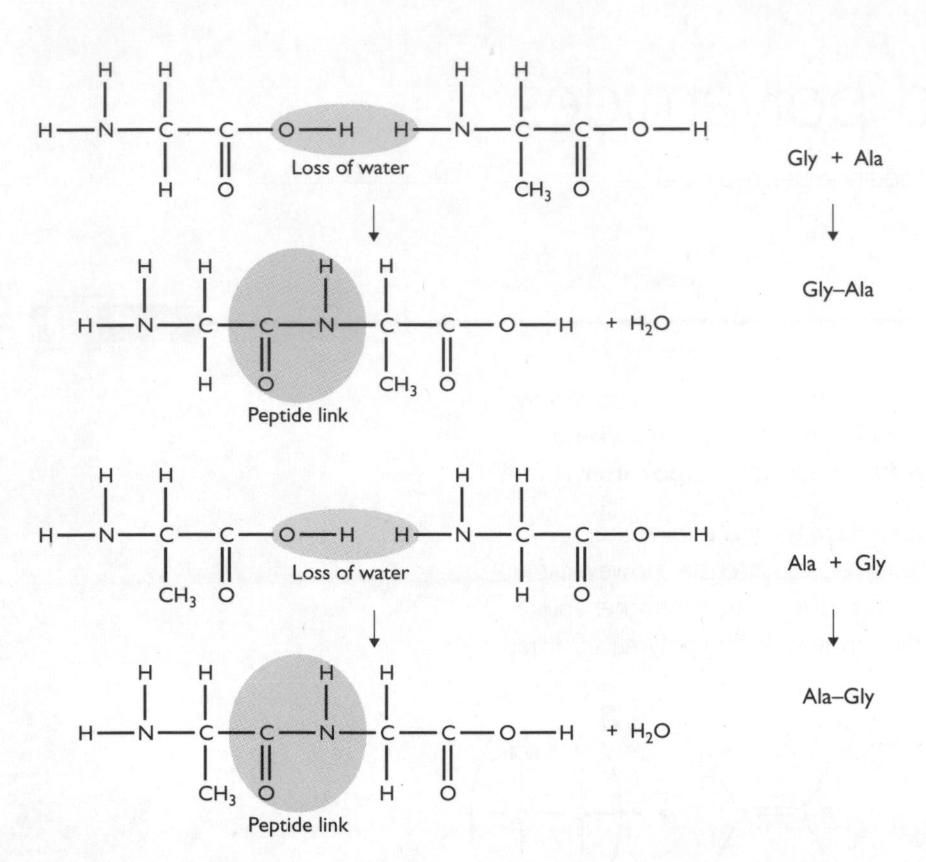

In a reaction mixture containing glycine and alanine, the dipeptides gly-gly and ala-ala would also be formed.

All dipeptides can react further with additional amino acids extending the chain length. This leads to the formation of polypeptides and proteins.

Peptides and proteins can be hydrolysed to the amino acids by refluxing the peptide (or protein) with 6.0 mol dm^{-3} hydrochloric acid or with aqueous sodium hydroxide solution.

Acid hydrolysis produces the cation; alkaline hydrolysis produces the anion.

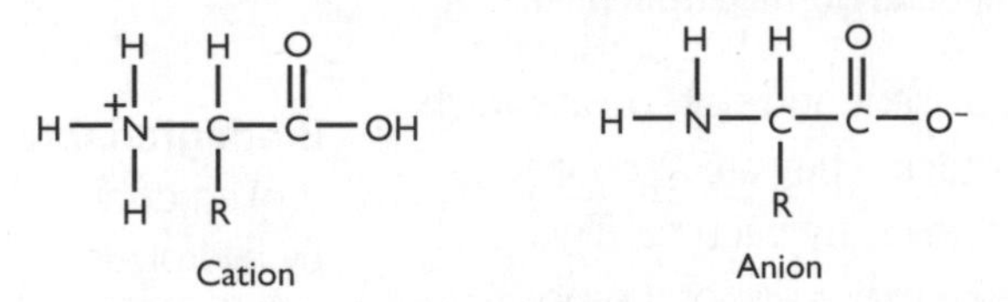

Typical mistake

In general, hydrolysis of proteins is well understood. However, candidates often lose marks by referring to amino acids being formed by hydrolysis when in fact cations (formed from acid hydrolysis) or anions (formed by alkaline hydrolysis) are produced.

Now test yourself

1 (a) Give the systematic names of the following amino acids:

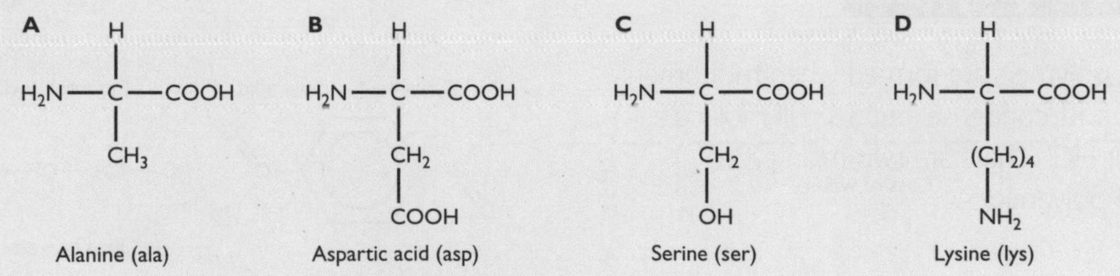

(b) Draw the two optical isomers of compound A.

(c) Draw the zwitterions formed by compounds A and C.

(d) Draw the ions formed by compounds B and D at pH 2 and at pH 12.

(e) Compounds A and C can form a mixture of dipeptides.
  (i) Identify how many different dipeptides can be formed.
  (ii) Draw these dipeptides.

Answers on p. 96

Polyesters and polyamides

There are two categories of polymers: addition polymers and condensation polymers.

Addition polymers

Revised

Alkenes can undergo addition reactions in which an alkene molecule joins others so that a long molecular chain is built up. The individual alkene molecule is a **monomer** and the long-chain molecule is a **polymer**.

Polymerisation can be initiated in a variety of ways and the initiator is often incorporated at the start of the long molecular chain. However, if the initiator is disregarded the empirical formulae of the monomer and the polymer are the same. Common monomers and the polymers formed from them are shown in Figure 2.1.

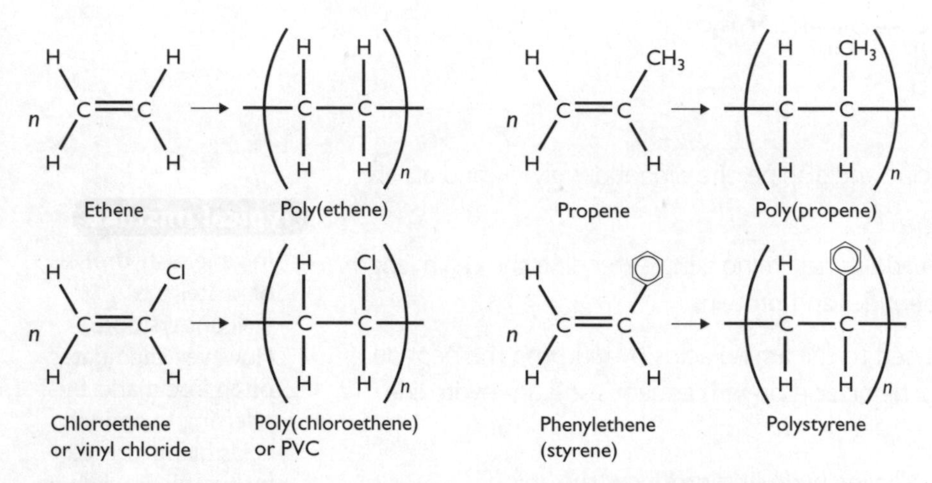

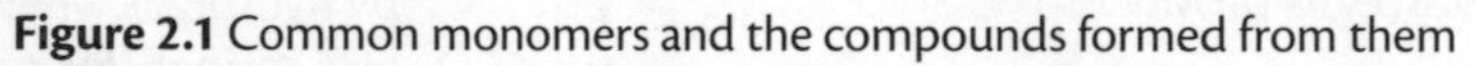

Figure 2.1 Common monomers and the compounds formed from them

The bonds in addition polymers are strong, covalent and non-polar, which makes most polymers resistant to chemical attack. They are also non-**biodegradable** because they are not broken down by bacteria. The widespread use of these polymers has created a major disposal problem.

Biodegradable polymers can be broken down either by bacteria or by hydrolysis.

Condensation polymers

Revised

Condensation polymers are formed when monomers react together and 'condense' out a small molecule such as H_2O or HCl. There are two main types: polyesters and polyamides.

Polyesters

Terylene® is a common polyester made by reacting the monomers ethene-1,2-diol and benzene-1,4-dicarboxylic acid.

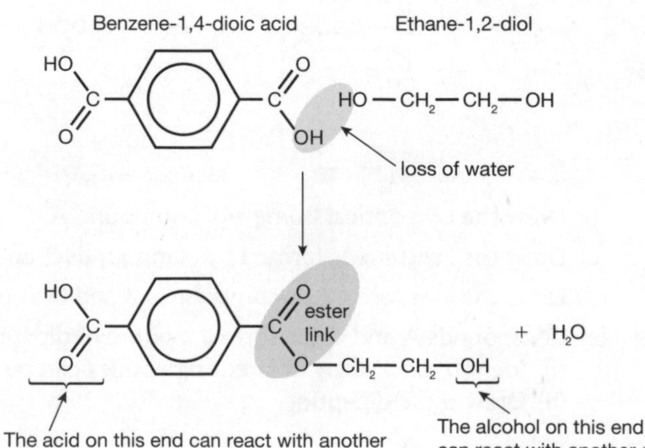

The resulting polymer is almost linear. This means that the polymer chains can be packed closely together. The close packing produces strong intermolecular forces that enable the polymer to be spun into thread.

Polyamides

Polyamides are prepared from two monomers, one with an amine group at each end and the other with a carboxylic acid group at each end. Nylon-6,6 is made from the two monomers 1,6-diaminohexane and hexane-1,6-dicarboxylic acid. It is called nylon-6,6 because each monomer contains six carbon atoms.

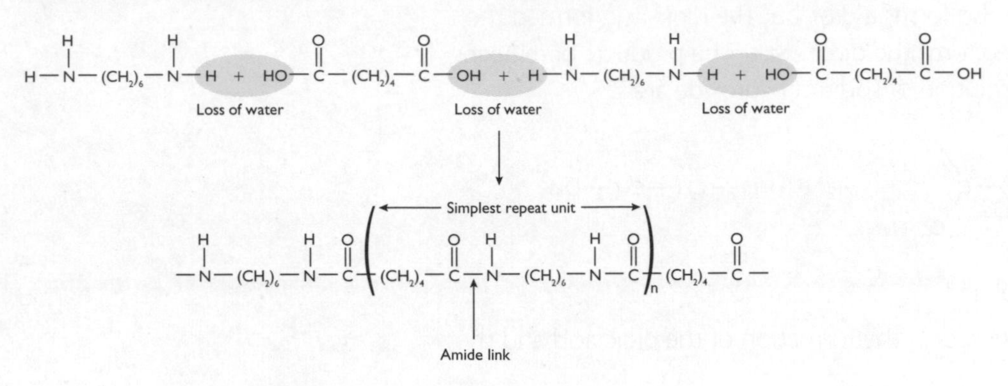

Figure 2.2 Formation of nylon-6,6

Nylon forms a strong flexible fibre when it is melt-spun.

Kevlar® is another polyamide. It is stronger than steel and fire resistant. It is used for making bulletproof vests, crash helmets and for protective clothing used by fire fighters. It is made from two monomers: benzene-1,4-diamine and benzene-1,4 dicarboxylic acid:

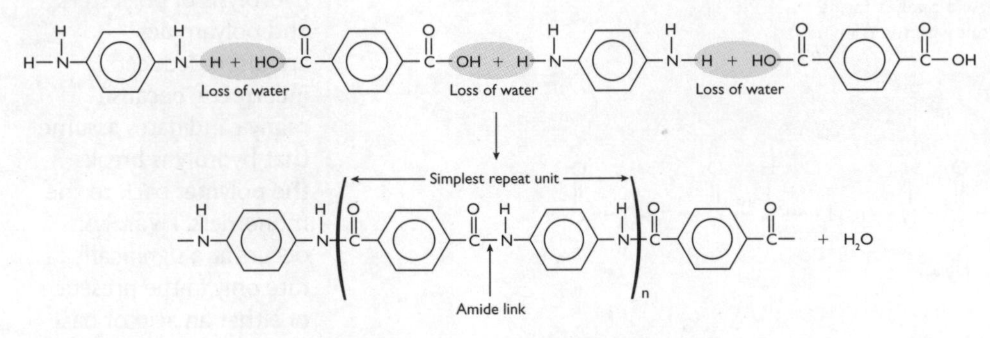

Figure 2.3 Formation of Kevlar®

Now test yourself

2 Draw two repeat units of the polymer that could be formed from:

(a) $HOH_2C—CH_2—CH_2—CH_2OH$ + $HO—C(=O)—(CH_2)_3—C(=O)—OH$

(b) $HO—C(=O)—CH_2—CH(CH_3)—NH_2$

Answers on p. 96

Hydrolysis and degradable polymers

Revised

The ester link in a polyester and the amide link in a polyamide are both polar links and are subject to acid-catalysed and base-catalysed hydrolysis.

Ester link

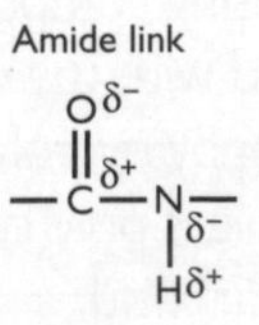

Acid hydrolysis of a polyester results in the formation of a diol and a dioic acid:

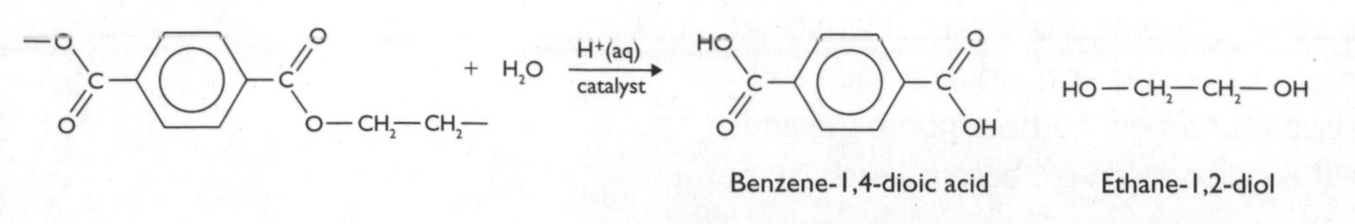

Base hydrolysis of a polyester also forms a diol, but the dioic acid formed then reacts with the base catalyst to form the dioate salt. The products of refluxing Terylene® with an aqueous solution of sodium hydroxide are:

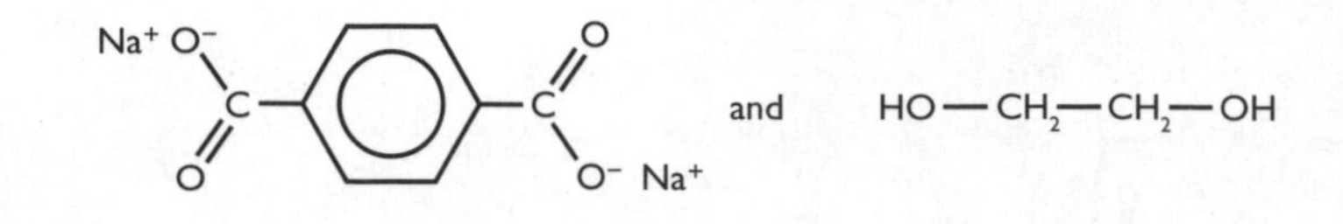

Polyamides can also be hydrolysed:

- Acid-catalysed hydrolysis results in the formation of the dioic acid and the di-salt of the diamine.
- Base-catalysed hydrolysis results in the formation of the diamine and the dioate salt of the dioic acid.

Hydrolysis of polyamides is summarised in the reaction scheme below.

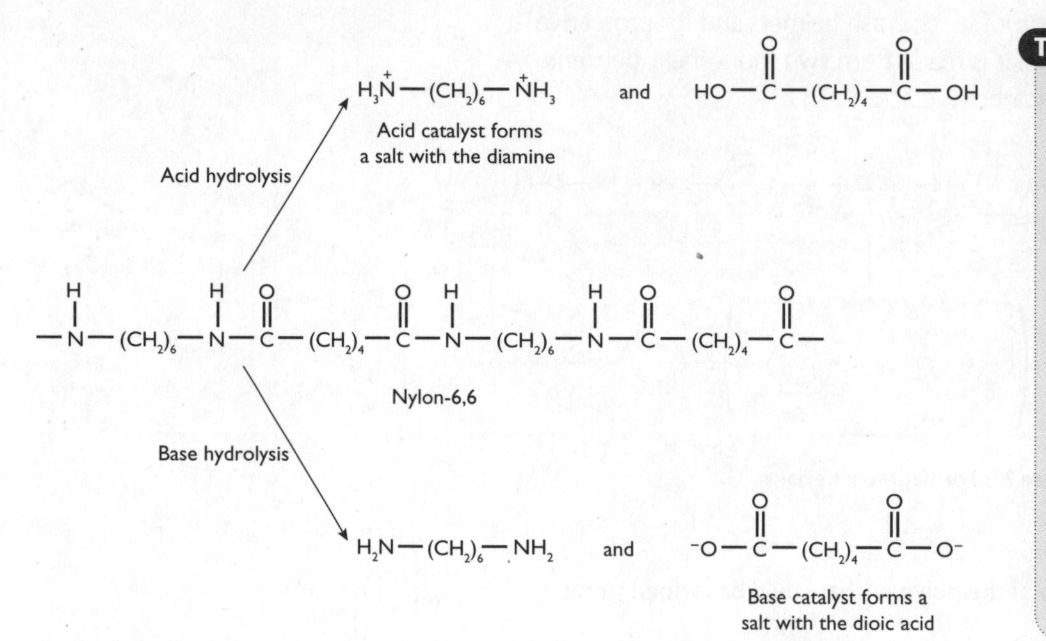

Typical mistake

The products of the hydrolysis of polyesters and polyamides are often deduced incorrectly because many candidates assume that hydrolysis breaks the polymer back to the monomers. Hydrolysis occurs at a significant rate only in the presence of either an acid or base catalyst. The presence of the catalyst results in the formation of the corresponding salts of the monomers.

Now test yourself

3 Identify the products formed by both acid and base hydrolysis of the polymer shown below:

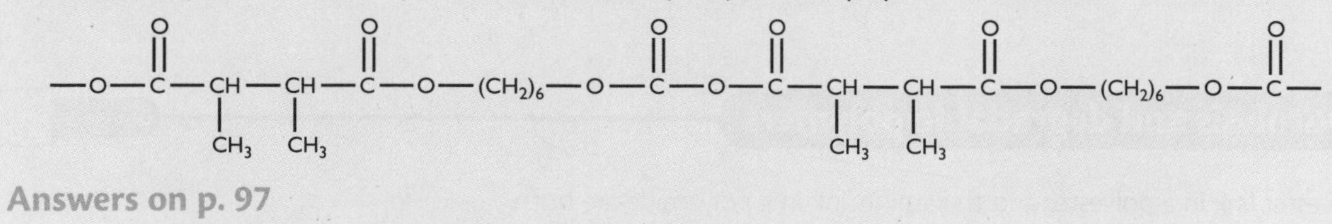

Answers on p. 97

Chemists are aware of the impact on the environment of using compounds derived from fossil fuels and also of the problems associated with the disposal of plastic waste. Condensation polymers can be disposed of by hydrolysis. Photodegradable polymers are being developed. An example is poly(lactic acid), PLA, which is prepared from 2-hydoxypropanoic (lactic) acid,

$CH_3CH(OH)COOH$. A molecule of lactic acid contains an alcohol group (–OH) and a carboxylic acid group (–COOH). These can react together to produce an ester linkage.

The introduction of polymers such as PLA reflects the role of chemists in minimising the impact on the environment through the use of renewable sources and by the development of degradable polymers. PLA is particularly attractive as a sustainable alternative to products derived from petrochemicals, since the monomer can be produced by the bacterial fermentation of agricultural by-products, such as cornstarch or sugar cane, which are **renewable** feedstocks. Polylactic acid is fully compostable and degrades to carbon dioxide and water in a relatively short period of time.

PLA is more expensive than many petroleum-derived commodity plastics, but its price has been falling as more production comes online. The degree to which the price will fall and the degree to which PLA will be able to compete with non-sustainable petroleum-derived polymers is uncertain.

Renewable relates to an essential chemical that can be replaced. An example is cornstarch, which is used in the production of PLA — more cornstarch can be produced by growing more corn.

Examiner's tip

Examiners are always looking for new ways to test routine chemistry and often set questions containing structures of complex molecules. Remember that, no matter how complex the molecule, only the functional groups react. A good starting point is to identify the functional groups within any complex molecule, as in the following example:

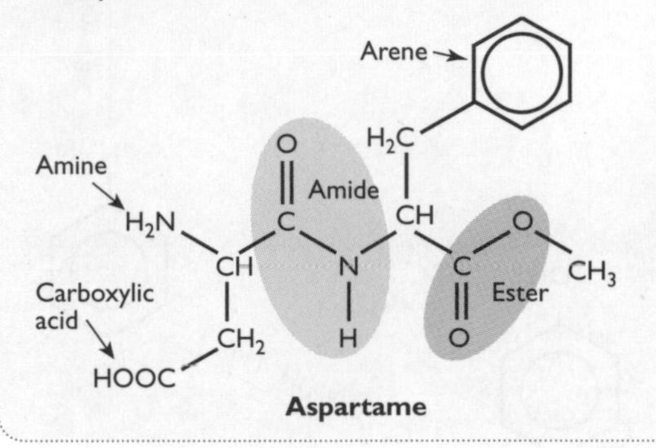

Although aspartame is a complex molecule, questions can only be asked about one or more of the functional groups: amide, amine, arene, ester or carboxylic acid.

Synthesis

Multistage conversions

Revised

It may not be possible to convert one chemical into another using a single reaction — intermediate compounds have to be formed. The flow charts below link together the functional groups covered so far in this unit and give the essential reagents and conditions. Flow chart 1 links together all the aliphatic chemistry.

Flow chart 1: Aliphatic chemistry

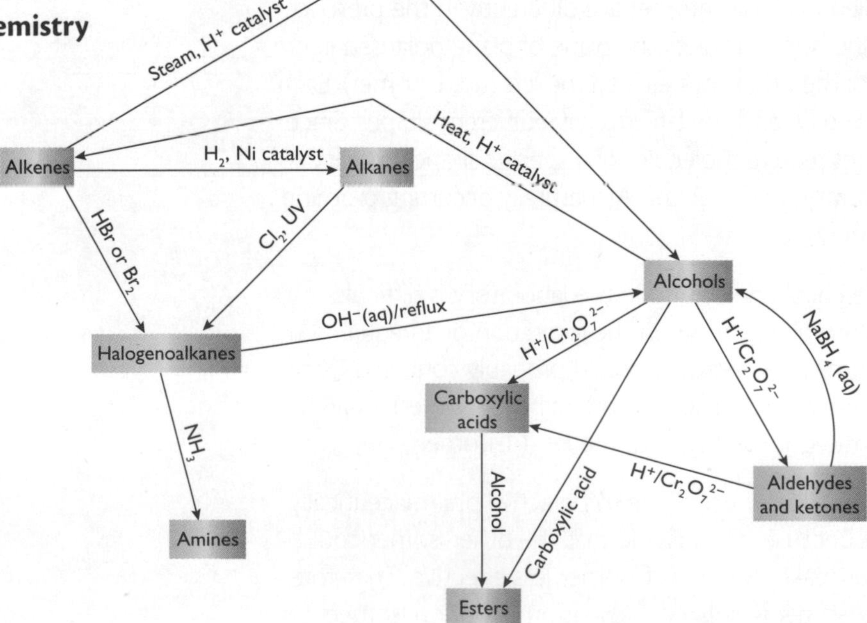

Flow chart 1 shows that an alkene can be converted into an alkane by a single step. However, to convert an alkane into an alkene, the alkane must first be converted into a halogenoalkane, then into an alcohol and finally into an alkene. This multistage conversion is less efficient than a single-stage conversion. You can use the flow chart to work out other multistage conversions.

Flow chart 2 is simpler and links together all the aromatic chemistry.

Flow chart 2: Aromatic chemistry

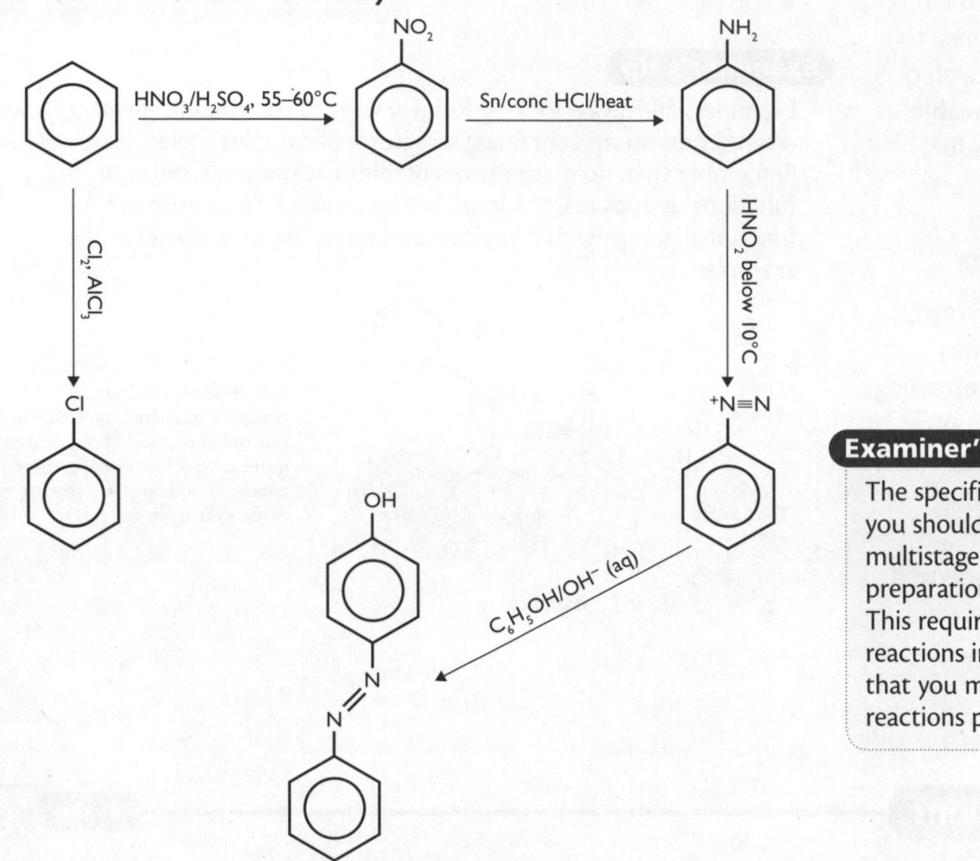

Examiner's tip

The specification outlines that you should be able to work out multistage syntheses for the preparation of organic compounds. This requires knowledge of all the reactions in the specification. Note that you may have to use unfamiliar reactions provided for you.

Chirality in pharmaceutical synthesis

Revised

Molecules with a chiral centre exist as one of two possible optical isomers. Optical isomers are so-called because they behave differently in the presence of plane-polarised light. One isomer rotates the plane of plane-polarised light to the right (the D-isomer); the other rotates it to the left (the L-isomer). Each isomer is optically active. The D- and the L-forms have different shapes. Since many biochemical processes require molecules of a specific shape it is easy to see why one isomer is naturally predominant. All naturally occurring α-amino acids occur in the L-form only.

The preparation of a single chiral compound in the laboratory is difficult. Laboratory synthesis would probably result in the formation of an equal amount of each optical isomer. The product would probably contain 50% of the isomer that rotates plane-polarised light to the right (D-isomer) and 50% of the isomer that rotates plane-polarised light to the left (L-isomer).

Only one of the isomers has the correct shape to be active pharmaceutically. It has to be separated out because it is possible that the other isomer could have adverse side effects or make the correct isomer less effective. Therefore, it is vital that the effective isomer is isolated. Separation of optical isomers can be achieved but it is tedious and expensive. Techniques such as chiral chromatography can be used.

The usual chromatographic techniques do not distinguish between optical isomers. However, if the column is packed with a solid that contains an active site in the form of either an enzyme or a chiral stationary phase, then one optical isomer will be adsorbed on to the column more effectively than the other and separation can be achieved.

By contrast, synthesis using naturally occurring enzymes or bacteria results in the formation of a single optical isomer. This occurs because active sites have specific shapes that promote specific reactions. Only the pharmaceutically active isomer is produced, which eliminates possible adverse side effects from the other optical isomer and also eliminates the need to separate the pharmacologically active isomer from the inactive isomer.

Single optical isomers can also be produced by using starting materials such as L-amino acids or L-sugars or by using a chiral catalyst. Chiral catalysts are now being used in the development of new drugs, the production of flavourings, sweetening agents and insecticides and in other aspects of material and medical science.

Check your understanding

1 (a) Nylon is not a single material but a group of chemicals made from diamines and dioic acids (or dioyl chlorides).

(i) Identify the monomers used to make:
- nylon-4,6
- nylon-6,4
- nylon-6,10

(ii) Different nylons have different melting points. Suggest why nylon-4,6 has a lower melting point than nylon-6,10.

(b) Nylon-6 can be made from the cyclic amide caprolactam, molecular formula $C_5H_{11}NO$.

(i) Draw the structure of caprolactam.

(ii) Draw two repeat units of nylon-6 formed from caprolactam.

2 Each of the following conversions involves multistage synthesis.

(a) propene ⟶ propanone

(b) 3-chloropropan-1-ol ⟶ 3-hydoxypropene

(c) phenylethanone ⟶ poly(phenylethene)

For each conversion:
- explain how it could be achieved
- write an equation and state the special conditions, if any, for each stage in the synthesis

3 During the late 1950s and early 1960s thalidomide, shown opposite, was given to pregnant women to combat morning sickness.

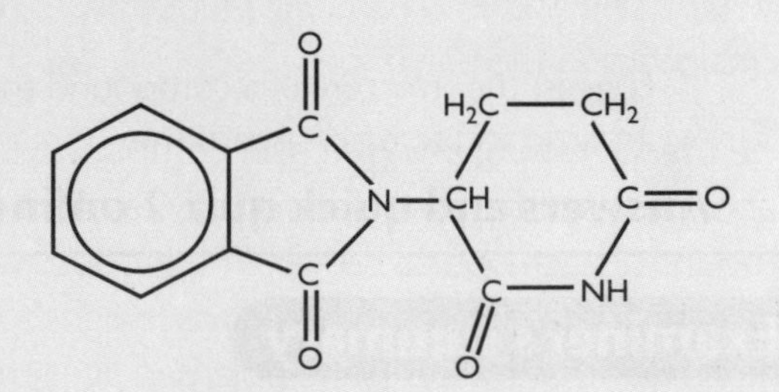

(a) What is the molecular formula of thalidomide?

(b) Copy the structure of thalidomide and identify the chiral carbon by drawing an asterisk next to it.

(c) Explain why it is important that pharmaceutical companies ensure that drugs contain a single optical isomer.

(d) Give two ways in which pharmaceutical companies can manufacture drugs that contain only a single optical isomer.

Answers on pp. 97, 98

Exam practice

1 Propene is an important industrial chemical essential in the production of a wide range of polymers, plastics and fibres. By far the greatest use of propene is as the monomer for polymerisation to poly(propene).

(a) (i) Draw the monomer propene. [1]

(ii) Draw a section of poly(propene) to show *two* repeat units. [2]

(b) There are difficulties caused by waste polymers such as poly(propene). Not only are they non-biodegradable but when burnt they produce a wide range of toxic fumes such as acrolein, $CH_2{=}CHCHO$, which has a choking odour and is the major cause of death of those suffocated in house fires. Identify the ***two*** functional groups present in acrolein and describe how you could test to show the presence of each group. Describe what you would see with each test. [6]

2 Consider the reaction scheme shown below.

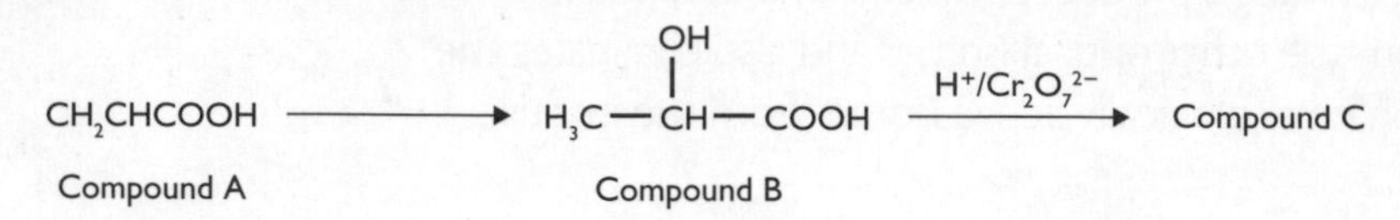

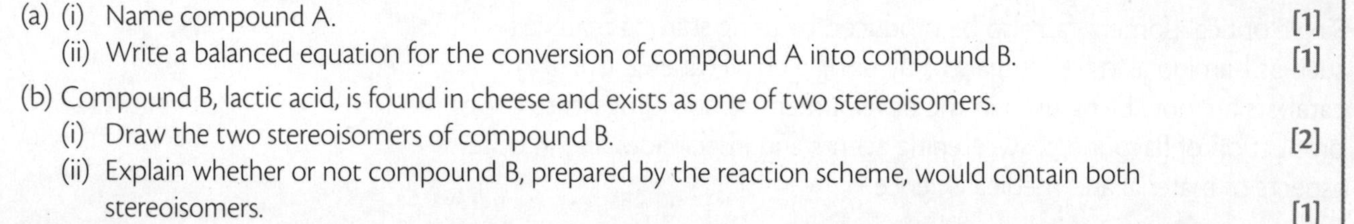

(a) (i) Name compound A. [1]

(ii) Write a balanced equation for the conversion of compound A into compound B. [1]

(b) Compound B, lactic acid, is found in cheese and exists as one of two stereoisomers.

(i) Draw the two stereoisomers of compound B. [2]

(ii) Explain whether or not compound B, prepared by the reaction scheme, would contain both stereoisomers. [1]

(c) (i) Draw the structure of compound C. [1]

(ii) Compound C can be reduced to propane-1,2-diol. Using [H] to represent the reducing agent, construct a balanced equation for this reduction. [3]

3 (a) Aspartame, shown below, is used as an artificial sweetener.

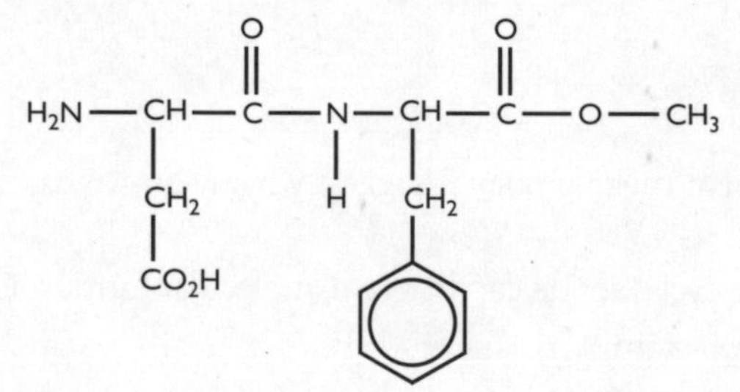

(i) Aspartame contains five functional groups including the benzene ring. Name the other *four* functional groups. [4]

(ii) *Two* of the four functional groups can be hydrolysed. Circle these groups on the diagram above. [2]

(iii) Show the structures of the organic products formed by the acid hydrolysis of aspartame. [3]

(b) (i) Aspartame has two chiral carbon atoms. Identify each with an asterisk (*). [2]

(ii) Explain what is meant by the term *chiral* and deduce the number of possible stereoisomers. [2]

(c) Aspartame can be made from aspartic acid (shown opposite).

O
H2N—CH—C—OH
CH2
CO2H

Suggest the structure of a compound that could react with aspartic acid to make aspartame. [2]

Answers and quick quiz 2 online

Online

Examiner's summary

You should now have an understanding of:

- ✔ amino acids, chirality and optical isomers
- ✔ peptide formation and the hydrolysis of proteins
- ✔ condensation polymers
- ✔ hydrolysis and degradable polymers
- ✔ synthetic routes
- ✔ chirality in pharmaceuticals

3 Analysis

Chromatography

Chromatography is an analytical technique that involves the small-scale separation of components within a mixture. All types of chromatography contain a **stationary phase** and a **mobile phase**.

Different types of chromatography separate the components in a mixture by either **adsorption** or by **partition**.

Separation by **partition** is achieved when solutes are not equally soluble in the mobile and stationary phases. An equilibrium is set up between the mobile and stationary phases.

Adsorption If the stationary phase is a solid, separation depends on the adsorption of each component onto the surface of the stationary phase and separation is achieved by adsorption.

Thin-layer chromatography (TLC)

Revised

The TLC plate is a glass, metal or plastic plate coated with a uniform thin layer of either silica gel, SiO_2, or alumina, Al_2O_3. The coating is the stationary phase. The mixture to be separated is spotted onto the base line (drawn in pencil) and is allowed to dry. The plate is then placed in a beaker containing the solvent (mobile phase). The beaker is then covered with a watch glass to ensure that the air inside the beaker is saturated with solvent vapour. This stops the solvent evaporating as it rises up the plate.

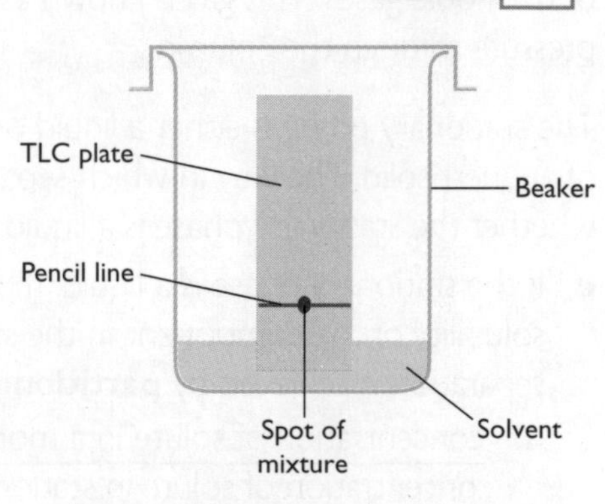

Figure 3.1 Apparatus for thin-layer chromatography

As the solvent travels up the plate, the components in the mixture interact with the surface of the stationary phase. This is called **adsorption**. The surface of the stationary phase is polar and hydrogen bonds, dipole–dipole interactions and van der Waals forces form between the polar surface and the components of the mixture. The greater the interaction, the greater is the adsorption and this restricts the distance travelled by that component.

Typical mistake

When describing how separation is achieved by thin-layer chromatography many students confuse 'adsorption' with 'absorption'. Separation is achieved because different molecules have different surface interactions with the stationary phase. The strength of the surface interactions determines how much a molecule adsorbs (sticks) to the surface of the stationary phase. Remember, it is 'ad...', *not* 'ab...'.

As the solvent rises up the plate the components in the mixture rise up with it. How fast each component rises depends on how soluble the component is in the solvent and how much the component adsorbs to the stationary phase. TLC achieves separation by adsorption.

The components in a mixture can be identified by using $\boldsymbol{R_f}$ **values**.

R_f value stands for 'retardation factor' and is measured by using the equation:

$$R_f = \frac{\text{Distance moved by spot/solute}}{\text{Distance moved by solvent}}$$

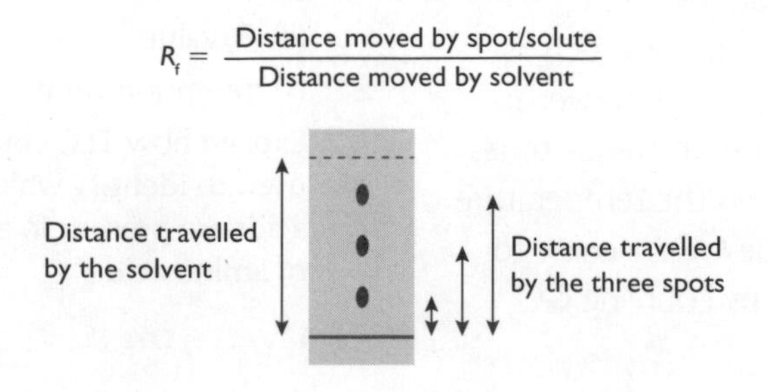

$\boldsymbol{R_f}$ **value** is the distance moved by a component divided by the distance moved by the solvent.

Gas chromatography (GC)

Revised

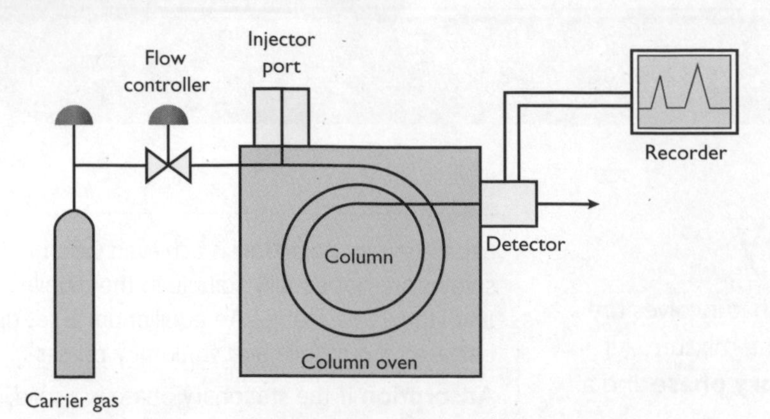

Figure 3.2 Diagram representing the apparatus for gas chromatography

Gas chromatography, GC, involves a sample being vaporised and injected into the chromatography column.

The mobile phase is an unreactive or inert gas such as nitrogen or one of the noble gases. This gas is known as the carrier gas and flows under pressure through the column.

The stationary phase is either a liquid or a solid adsorbed onto the surface of an inert solid. The way in which separation is achieved depends on whether the stationary phase is a liquid or a solid:

- If the stationary phase is a liquid, separation depends on the relative solubility of the component in the stationary and mobile phases and separation is achieved by **partition**.

$$\frac{\text{concentration of solute in a mobile phase}}{\text{concentration of solute in stationary phase}} = \text{constant}$$

- If the stationary phase is a solid, separation depends on the adsorption of the component onto the surface of the stationary phase and separation is achieved by **adsorption**.

The time between the injection of the sample and the emergence of a component from the column is called the **retention time**. Retention time depends on the volatility of the solute and the relative solubility of the solutes in the mobile and stationary phases.

Retention time is the time taken from the injection of the sample for a component to leave the column.

A detector is used to monitor the outlet stream from the column in order to determine both the time at which each component reaches the outlet and the amount of each component. The recorder produces a chromatogram showing each component as a separate peak. The substances are identified qualitatively by the order in which they emerge from the column and quantitatively by the area of the peak of each component.

Analysis by GC has its limitations, in that similar compounds often have similar retention times. It is likely that in a mixture of gases such as methane, ethane, propane and butane some of the peaks would overlap. Identification of unknown compounds is difficult because reference times vary depending on the flow rate of the carrier gas and on the temperature of the column — indeed, retention times vary from one GC machine to another. These limitations have been largely overcome by coupling GC with mass spectrometry.

Now test yourself

1 Define each of the following terms:
(a) stationary phase
(b) mobile phase
(c) adsorption
(d) partition
(e) R_f value
(f) retention time

2 Explain how TLC could be used to identify which amino acids are present in a mixture of amino acids.

Answers on p. 99

GC–MS

Revised

The combination of gas chromatography and mass spectrometry provides a powerful analytical tool.

The components in a mixture are separated by gas chromatography and then each component is analysed separately by mass spectrometry. The component is vaporised and then ionised. The process of fragmentation was outlined in Unit F322 and a number of simple fragment ions were identified in order to determine the relative isotopic masses of elements. The mass spectrum of propane is shown in Figure 3.3.

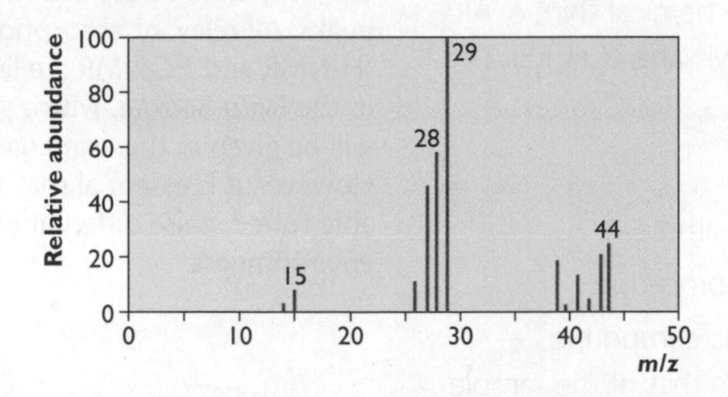

Figure 3.3 Mass spectrum of propane

The relative molecular mass of propane is 44 and the peak furthest to the right of the spectrum represents the ion $C_3H_8^+$. This is called the **molecular ion peak**. However, the bombardment of electrons also creates ions of fragments of the molecule that also register on the printout.

Even in molecules as simple as propane there are a large number of lines in the mass spectrum. The combination of the lines and their size is specific to the individual compound — the fragmentation peaks are said to be a **fingerprint** of the molecule. This fingerprint can be cross-matched against a computer database to identify the compound.

Coupling together GC and MS enables separation of components in a mixture (GC) and identification of each component (MS). This provides a strong analytical tool, which is used widely in many aspects of life — for example forensics, environmental analysis, airport security, food and drink analysis, and in medical applications.

Examiner's tip

Remember that it is a positive gaseous ion that is detected in a mass spectrometer. If, for example, you are asked to identify the ion responsible for the peak at *m/z* = 15, the answer is CH_3^+(g), *not* CH_3^+ or CH_3(g) or CH_3.

Now test yourself

3 State four uses of GC–MS.

Answers on p. 99

Spectroscopy

NMR spectroscopy

NMR spectroscopy involves the interaction of nuclei with radio waves, which are at the low-energy end of the electromagnetic spectrum.

If the nucleus of an atom contains an odd number of protons and/or neutrons, the nucleus has a net nuclear spin that can be detected by using radio frequency — for example, ^{1}H and ^{13}C can both be detected.

The nucleus behaves like a tiny bar magnet and as it spins it generates a magnetic moment. Adjacent nuclei also have magnetic moments.

Therefore, each nucleus is affected by neighbouring nuclei. The frequency at which each nucleus absorbs radio waves depends on its environment.

If the nucleus of an atom contains an even number of protons and an even number of neutrons the nucleus does *not* have a net nuclear spin and *cannot* be detected by using radio frequency — for example, ^{12}C and ^{16}O cannot be detected.

^{1}H and ^{13}C both absorb energy in the radio-wave part of the spectrum. However, the frequency of the radio waves absorbed depends on the surrounding atoms, i.e. the exact frequency absorbed depends on the chemical environment. This variation in the frequency absorbed is the key to the determination of structure. It is known as the chemical shift, δ. All absorptions are measured relative to TMS, tetramethylsilane, $(CH_3)_4Si$. The chemical shift, δ, of TMS is set at zero.

TMS is used as a standard because:

- it is chemically inert and does not react with the sample
- it is volatile and easy to remove at the end of the procedure
- it absorbs at a higher frequency than other organic compounds — therefore, its mass spectrum does not overlap with that of the sample

Examiner's tip

You do not have to learn the chemical shift values of any of these peaks. All relevant absorptions for ^{1}H-NMR and ^{13}C-NMR are listed in the *Data Booklet*, which you will be given in the examination. However, it is essential that you are able to recognise different chemical environments.

^{13}C-NMR spectroscopy

Revised

Carbon-12, ^{12}C, is the most abundant isotope of carbon. It does not have spin because it has an even number of protons and an even number of neutrons. The second isotope of carbon, carbon-13, ^{13}C, can be detected using low-energy radio waves and it is possible to generate ^{13}C-NMR spectra. The ^{13}C atom is about 6000 times more difficult to detect than ^{1}H atoms because of its low abundance (only about 1.1% of naturally occurring carbon is ^{13}C) and its low magnetic moment. Interaction between adjacent ^{13}C atoms is unlikely because of their low abundance. ^{13}C atoms do interact with adjacent protons but these interactions are removed by decoupling and all absorptions appear as singlets. Each peak represents a different carbon environment.

Ethanol has two carbons atoms, C_1 and C_2.

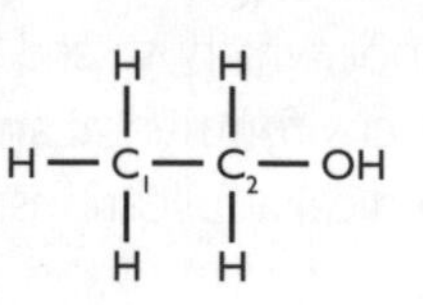

Therefore, there are two separate peaks, one for each carbon environment.

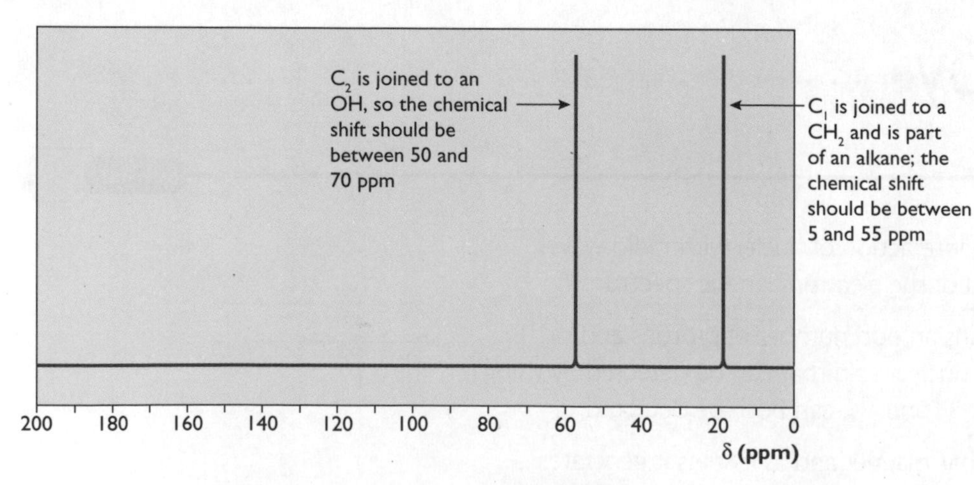

Figure 3.4 ^{13}C-NMR spectrum of ethanol

The key to interpreting ^{13}C-NMR spectra is to identify the number of different carbon environments and then to match them with the groups in the data sheet.

Example

Determine the number of carbon environments in propan-2-ol. For each environment predict the chemical shift, δ.

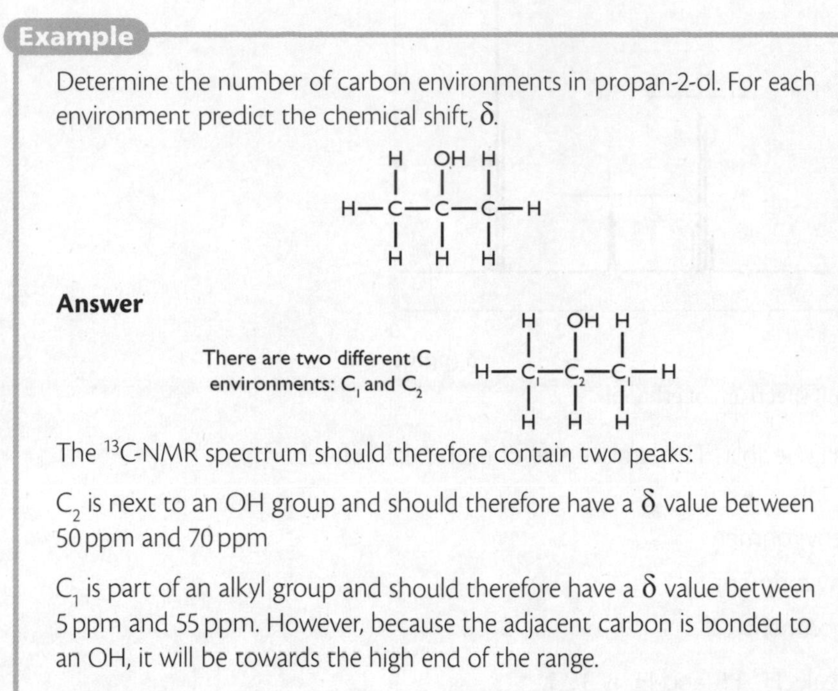

Answer

There are two different C environments: C_1 and C_2

The ^{13}C-NMR spectrum should therefore contain two peaks:

C_2 is next to an OH group and should therefore have a δ value between 50 ppm and 70 ppm

C_1 is part of an alkyl group and should therefore have a δ value between 5 ppm and 55 ppm. However, because the adjacent carbon is bonded to an OH, it will be towards the high end of the range.

The ^{13}C spectrum of propan-2-ol is shown below.

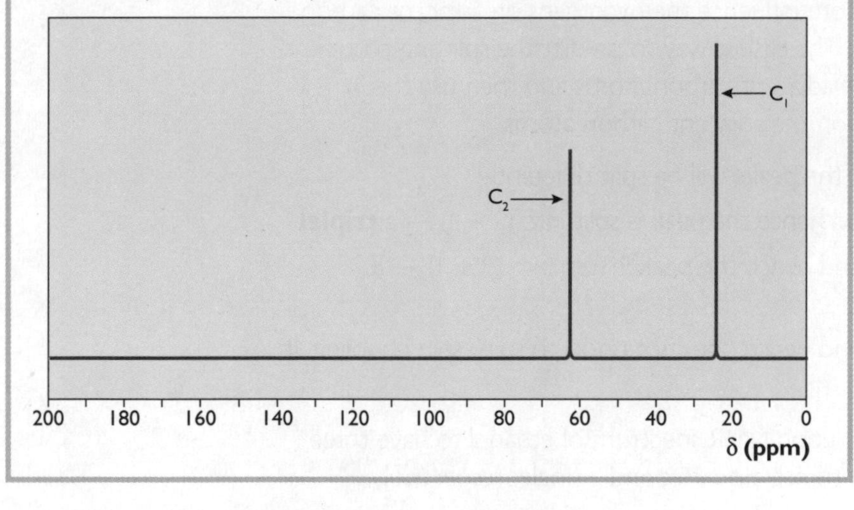

^{1}H (proton) NMR spectroscopy

Revised

Ethanol, C_2H_5OH, contains six hydrogen atoms, but they are not all identical.

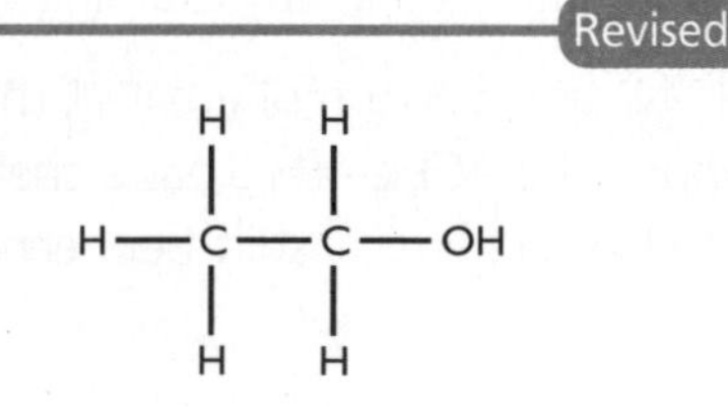

- The three hydrogens in the CH_3 group are all in the same environment and are labelled H_a.
- The two hydrogens in the middle CH_2 group are in the same environment and are labelled H_b.
- The hydrogen in the OH group is different from all of the rest and is labelled H_c.

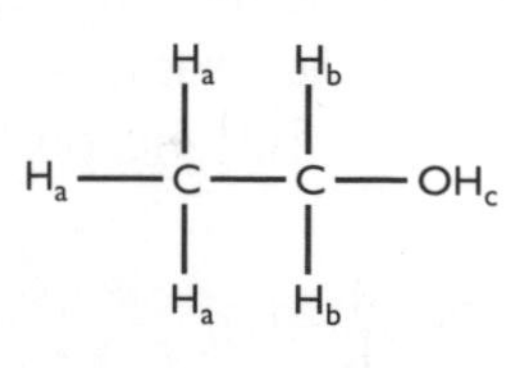

So the six hydrogens in ethanol are in three different environments. Therefore, the ^{1}H-NMR spectrum of ethanol contains three different peaks (H_a, H_b and H_c) with different chemical shifts.

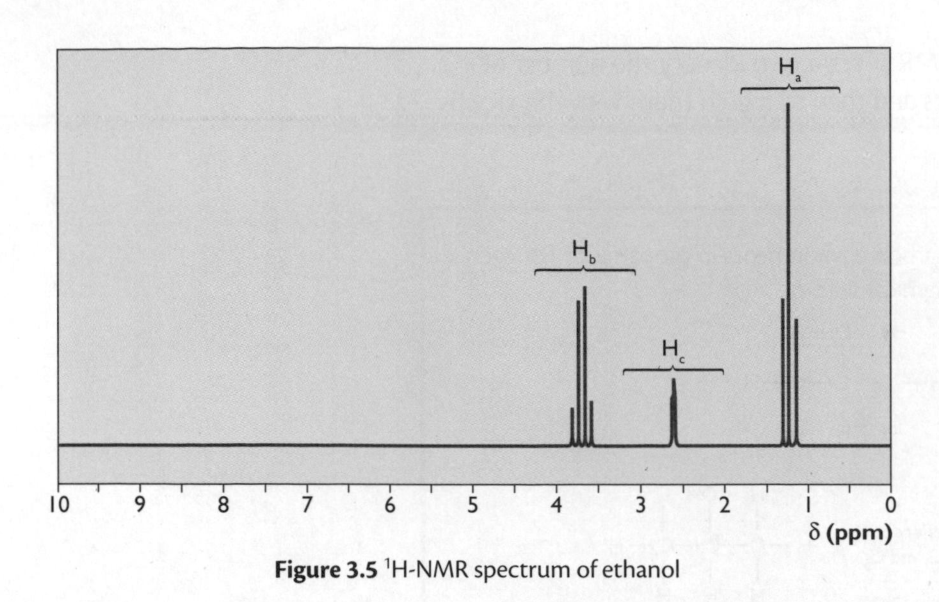

Figure 3.5 ^{1}H-NMR spectrum of ethanol

The three peaks are of different sizes and may be split. The relative size of each peak reflects the number of hydrogens in each environment:

- H_a — there are three hydrogens in this environment
- H_b — there are two hydrogens in this environment
- H_c — there is one hydrogen in this environment

It follows that the relative intensity of the peaks H_a, H_b and H_c is 3:2:1.

The hydrogens attached to one carbon atom influence the hydrogens on adjacent carbon atoms. This is called **spin–spin coupling**. The easiest way to predict the splitting pattern is to count the number of hydrogens on the adjacent carbon atoms and then use the **'*n* + 1'** rule, where *n* is the number of hydrogens on the adjacent carbon atoms.

In the NMR spectrum of ethanol, each of the peaks will be split differently.

- H_a is next to two hydrogens in CH_2 and hence the peak is split into (2 + 1) — a **triplet**
- H_b is next to three hydrogens in CH_3 and hence the peak is split into (3 + 1) — a **quartet**
- H_c is not attached to a carbon atom and hence does not undergo spin–spin coupling. It is, therefore, a **singlet**.

We would, therefore, expect the high-resolution NMR spectrum of ethanol to have three peaks of relative intensity 3:2:1, split into a triplet, a quartet and a singlet respectively.

The exact position (the chemical shift) of each peak can be obtained from the data sheet which will be supplied in any examination.

In the ^{1}H-NMR spectrum of ethanol, the $-CH_3$ (H_a) has a chemical shift between 0.7 ppm and 2.0 ppm, the $-CH_2-$ (H_b) has a chemical shift between 3.2 ppm and 4.3 ppm and the $-OH$ (H_c) has a chemical shift between 1.0 ppm and 5.5 ppm (see Figure 3.5).

Use of D_2O

The O–**H** and the N–**H** peaks have chemical shifts that differ between compounds and sometimes lie outside the range 1.0–5.5 ppm. Therefore, they are difficult to assign. When alcohols, carboxylic acids or amines are dissolved in water there is a rapid exchange between the protons in the functional groups (**labile protons**) and the protons in the water. For example:

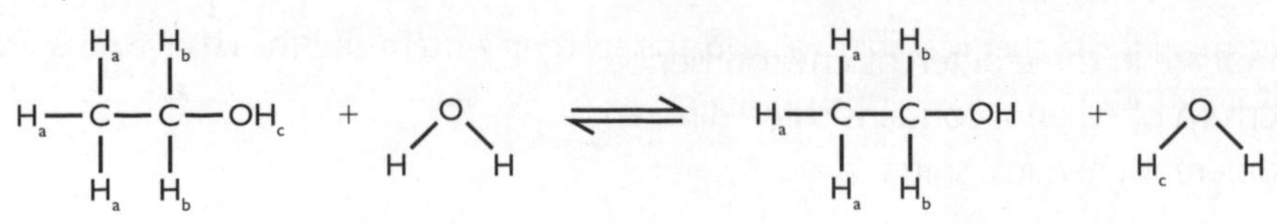

If water is replaced by deuterated water, ^2H_2O, the peak at H_c disappears. The H_c proton is replaced by deuterium, ^2H, which does not absorb in this region of the spectrum.

Deuterated water, ^2H_2O can be written as D_2O. Ethanol dissolved in deuterated water can be represented as:

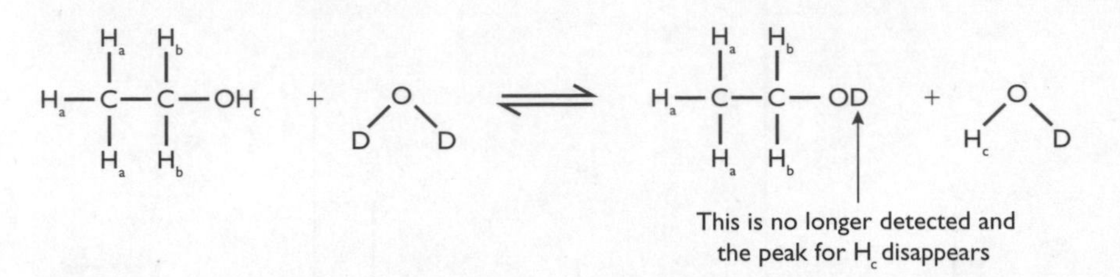

The use of ^2H_2O to identify labile protons is a valuable technique in proton ^1H-NMR.

When samples are prepared for NMR it may be necessary to dissolve them in a suitable solvent. Solvents containing protons are unsuitable because the protons would be detected and interfere with the spectrum. This is overcome by using a deuterated solvent, such as $CDCl_3$.

Combined techniques

Revised

Analytical chemistry is rather like detective work — pieces of evidence are gathered from different places. Usually no single piece of evidence is conclusive but when the pieces of evidence are slotted together, the combination is definitive. Gathering together the evidence is a bit like doing a jigsaw.

Examiner's tip

You may be faced with three or four different spectra that you have to use to identify a molecule. A good simple approach is as follows:

- **Start with the infrared spectrum and check for absorptions due to C=O and/or O–H**
- **Use the mass spectrum to identify the molecular ion and hence the molar mass.**
- **The number of peaks in the ^{13}C-NMR spectrum indicates the number of different carbon environments.**
- **The number of peaks in the ^1H-NMR spectrum indicates the number of different hydrogen environments.**
- **Use the splitting patterns in the ^1H-NMR spectrum to determine the adjacent hydrogen environments.**

Check your understanding

1 Cyclopentanone, C_5H_8O, is shown in the diagram.

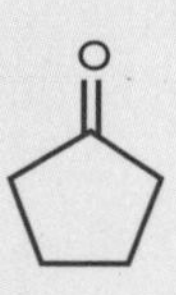

(a) Explain why the ^1H-NMR spectrum of cyclopentanone has two peaks and the ^{13}C-NMR spectrum has three peaks.

(b) Using the *Data Sheet*, predict the value of the chemical shift, δ, for each peak in the ^{13}C-NMR spectrum.

(c) Using the *Data Sheet*, suggest the chemical shift value, δ, the splitting pattern and the relative peak areas for each peak in the ^1H-NMR spectrum.

2 Identify compound X from the spectra below.

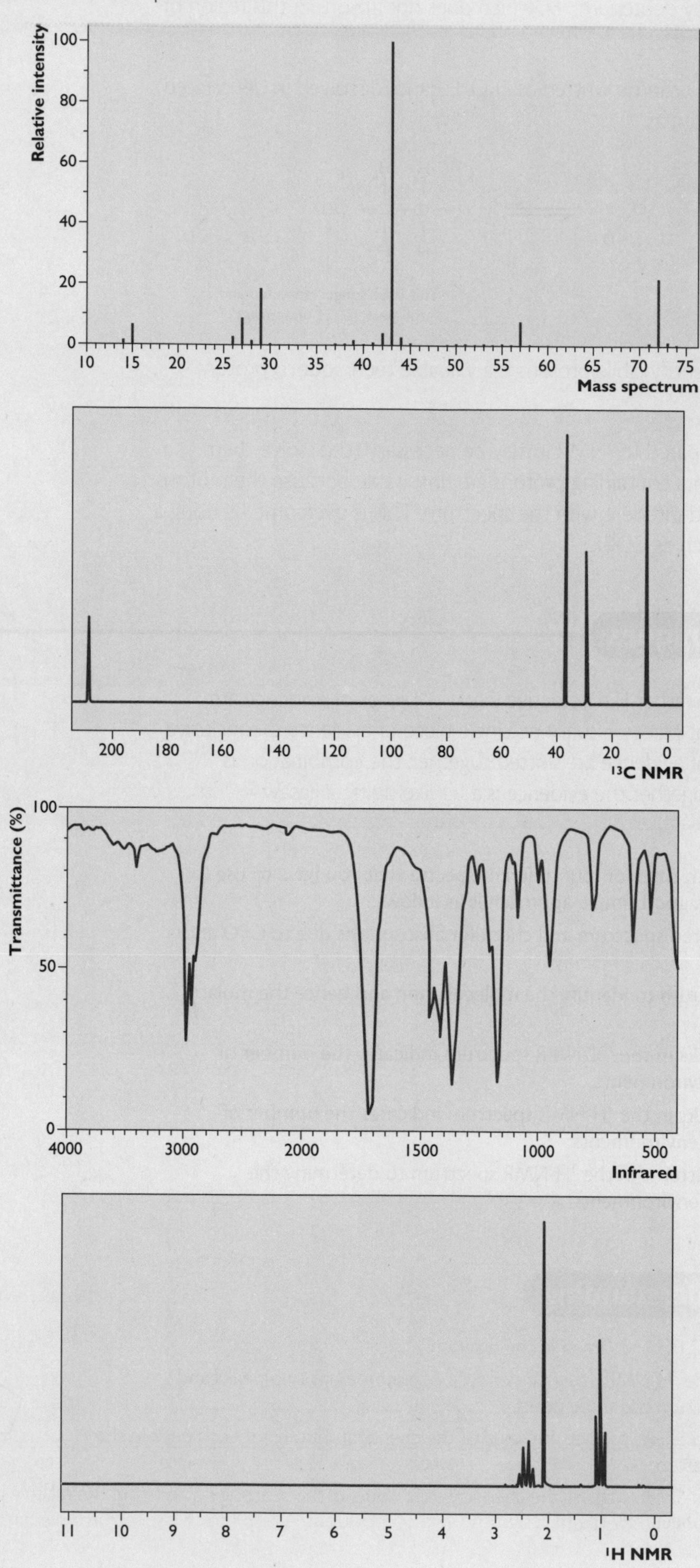

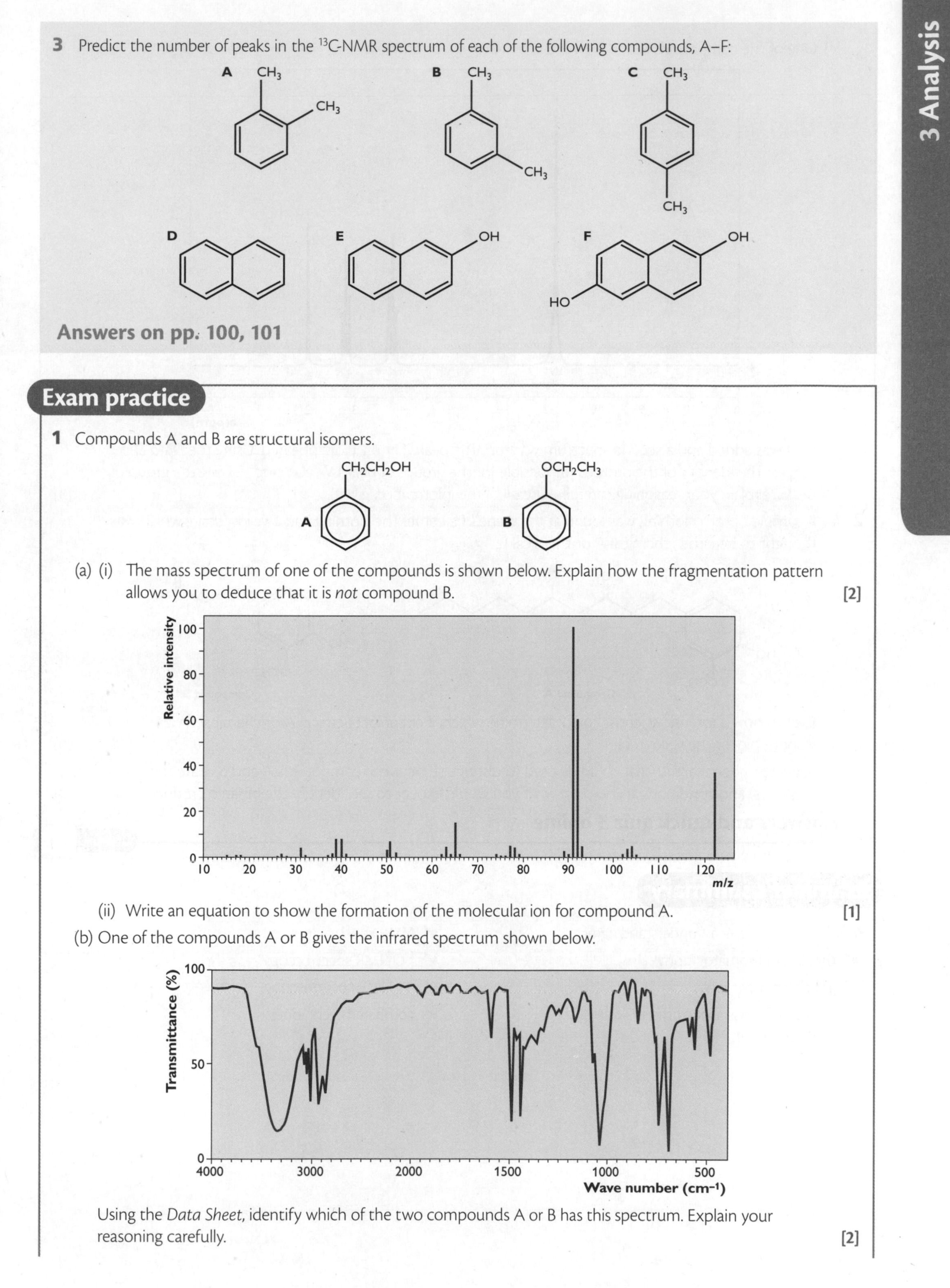

3 Predict the number of peaks in the ^{13}C-NMR spectrum of each of the following compounds, A–F:

Answers on pp. 100, 101

Exam practice

1 Compounds A and B are structural isomers.

(a) (i) The mass spectrum of one of the compounds is shown below. Explain how the fragmentation pattern allows you to deduce that it is *not* compound B. [2]

(ii) Write an equation to show the formation of the molecular ion for compound A. [1]

(b) One of the compounds A or B gives the infrared spectrum shown below.

Using the *Data Sheet*, identify which of the two compounds A or B has this spectrum. Explain your reasoning carefully. [2]

(c) One of the compounds A and B gives the ^{1}H-NMR spectrum below.

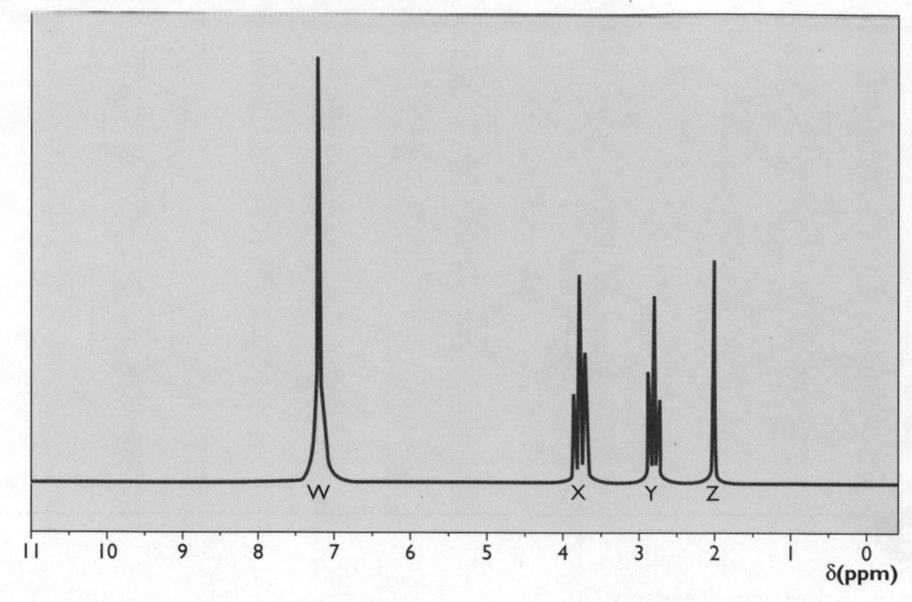

D_2O was added and a second spectrum was run. The peak Z at δ 2.0 disappeared. Using the *Data Sheet*, suggest the identity of the protons responsible for the groups of peaks W, X, Y and Z. For each group of peaks, explain your reasoning carefully. Use *all* of the information given. [8]

2 (a) A sample of torn clothing was found at the scene of a crime. The clothing had a yellow stain, which was thought to be either chemical A or chemical B:

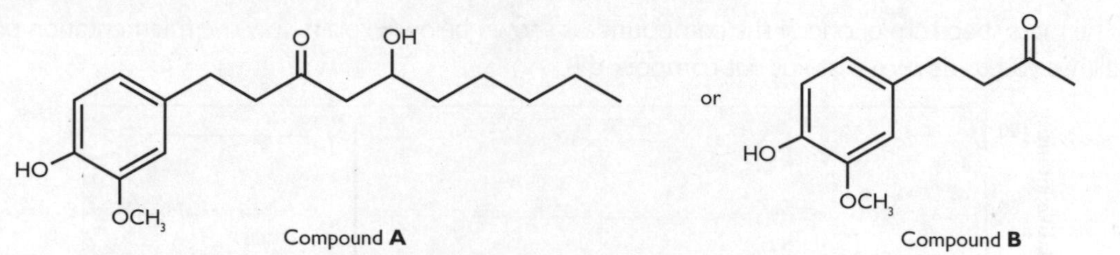

Explain how a forensic scientist could determine which, if either, of chemical A or chemical B was responsible for the yellow stain. [5]

(b) Suggest a chemical test that could be used to distinguish between compounds A and B. State the reagent(s) and conditions and explain what you would expect to see. Identify the organic product. [3]

Answers and quick quiz 3 online

Online

Examiner's summary

You should now have an understanding of:

- ✔ thin-layer chromatography
- ✔ gas chromatography
- ✔ combining mass spectrometry with gas chromatography
- ✔ NMR spectroscopy
- ✔ ^{13}C-NMR spectroscopy
- ✔ ^{1}H-NMR spectroscopy
- ✔ combined techniques

4 Rates, equilibria and pH

How fast?

This topic builds on your understanding of the AS reaction rates chemistry. It involves measuring and calculating reaction rates using rate equations.

Reaction rate

Revised

The rate of a reaction is usually measured as the change in concentration of a reaction species with time. The units of rate are mol dm^{-3} s^{-1}.

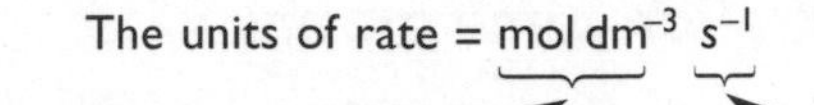

Measuring rates from graphs

Consider the reaction:

$$A + B \rightarrow C + D$$

It is possible to measure the rate of disappearance of either A or B or the appearance of one of the products, C or D. The data can be plotted on a concentration–time graph. The gradient of the tangent is a measure of the reaction rate.

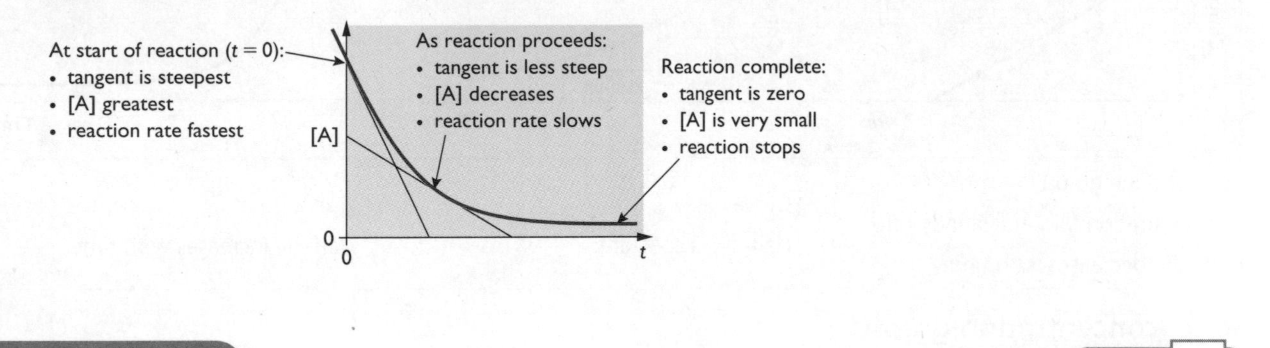

The rate equation

Revised

Orders of reaction

The rate equation of a reaction shows how the rate is affected by the concentration of each reactant. It can only be determined by experiment.

In general, for a reaction: A + B → C + D, the reaction rate is given by:

$$rate = k[A]^m[B]^n$$

- k is the rate constant of the reaction.
- m and n are the **orders of reaction** with respect to **A** and **B**.

The overall order of reaction is $(m + n)$.

You know from Unit F322 that increasing concentration usually results in an increased rate of reaction. However, different reagents can behave in a different manner. If we double the concentration of a reagent and the rate increases proportionately (i.e. the rate also doubles) then the reaction is said to be **first order** with respect to that reagent. If doubling the concentration of a reagent results in a four-fold increase in reaction rate, the reaction is said to be **second order** with respect to that reagent. If increasing the concentration of a reagent has no effect, the reaction is said to be **zero order** with respect to that reagent.

The rate constant

The rate constant k indicates the rate of the reaction:

- A large value of k means a fast rate of reaction.
- A small value of k means a slow rate of reaction.

An increase in temperature speeds up the rate of most reactions by increasing the rate constant k.

Determination of orders from graphs

Revised

Concentration–time graphs

During a reaction the concentrations of the reagent(s) and product(s) change. Measuring the concentration of a reactant at regular time intervals produces data that can be plotted on a graph. The shape of the resultant graph can be used to predict the order of reaction. This is determined by measuring the **half-life** of a reactant. The half-life of a reactant is the time taken for its concentration to halve during the reaction.

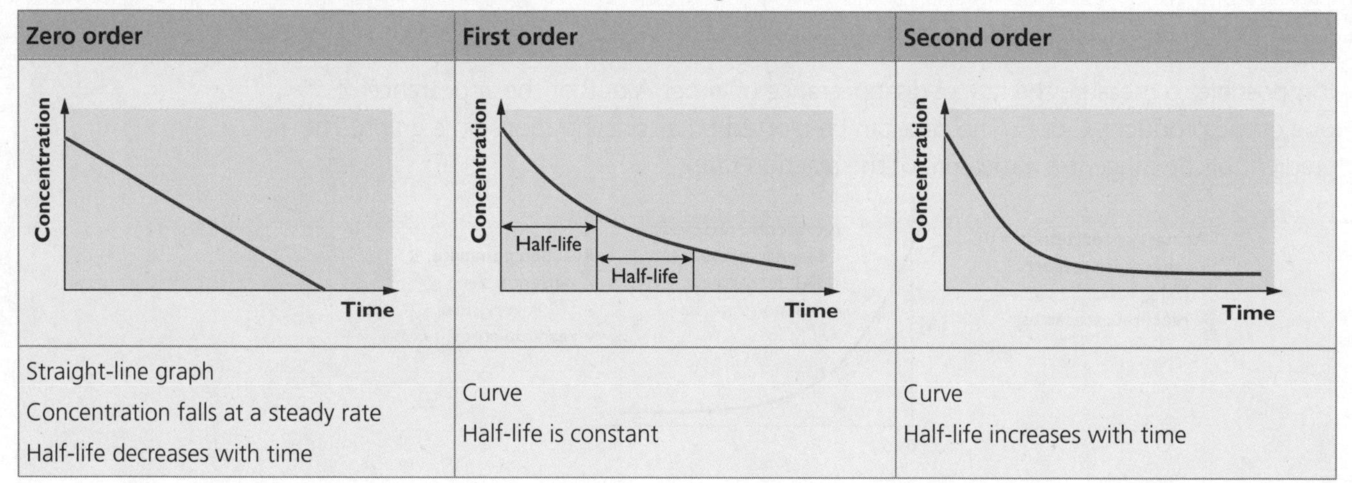

Zero order	First order	Second order
Concentration / Time graph	Concentration / Time graph (Half-life, Half-life)	Concentration / Time graph
Straight-line graph Concentration falls at a steady rate Half-life decreases with time	Curve Half-life is constant	Curve Half-life increases with time

Rate–concentration graphs

A concentration–time graph is first plotted and tangents are drawn at several time values on it, giving values of reaction rates. A second graph is now plotted of rate against concentration.

Zero order	First order	Second order
Rate / Concentration graph	Rate / Concentration graph	Rate / Concentration graph
rate $\propto [X]^0$ rate = constant Rate unaffected by changes in concentration	rate $\propto [X]^1$ Rate doubles as concentration doubles	rate $\propto [X]^2$ Rate quadruples as concentration doubles

Measuring rates using the initial rates method

Revised

For a reaction, A + B → C + D experiments are carried out using different initial concentrations of the reactants A and B.

When changing only one variable at a time, three experiments are required:

- In experiments 1 and 2, the concentration of A is changed and the concentration of B is kept constant.
- In experiments 2 and 3, the concentration of B is changed and the concentration of A is kept constant.

A typical set of results is shown in the table below.

Experiment	[A(aq)]/mol dm^{-3}	[B(aq)]/mol dm^{-3}	Initial rate/mol dm^{-3} s^{-1}
1	1.0×10^{-2}	1.0×10^{-2}	4.0×10^{-3}
2	2.0×10^{-2}	1.0×10^{-2}	1.6×10^{-2}
3	2.0×10^{-2}	2.0×10^{-2}	3.2×10^{-2}

Order of reaction with respect to each reagent

Order of reaction with respect to A

Comparing experiments 1 and 2: [B(aq)] is constant and [A(aq)] is doubled. The rate has quadrupled. Therefore, the reaction is second order with respect to **A**(aq):

$$\text{rate} \propto [A(aq)]^2$$

Order of reaction with respect to B

Comparing experiments 2 and 3: [A(aq)] is constant and [B(aq)] is doubled. The rate has doubled. Therefore, the reaction is first order with respect to B(aq):

$$\text{rate} \propto [B(aq)]$$

The rate equation

Combining the two orders with respect to the two reagents gives:

$$\text{rate} \propto [A(aq)]^2[B(aq)] \text{ or rate} = k[A]^2[B]$$

The overall order of this reaction is (2 + 1) = third order.

The rate constant, k, can be determined by rearranging the rate equation:

$$k = \frac{\text{rate}}{[A]^2[B]}$$

Substituting values from experiment 1 gives:

$$k = \frac{(4.0 \times 10^{-3})}{(1.0 \times 10^{-2})^2(1.0 \times 10^{-2})} = 4.0 \times 10^3 \text{ dm}^6 \text{ mol}^{-2} \text{ s}^{-1}$$

Units of rate constants

The units of a rate constant depend upon the rate equation for the reaction.

Order	Rate equation	Units of k
First	rate = k[A]	s^{-1}
Second	rate = k[A]^2	mol^{-1} dm^3 s^{-1} *or* dm^3 mol^{-1} s^{-1}
Third	rate = k[A]^2[B] or k[A]^3	mol^{-2} dm^6 s^{-1} *or* dm^6 mol^{-2} s^{-1}

Now test yourself

1 The rate equation for a reaction is:

rate = k[A][B]^2[C]

(a) What is the overall order of reaction and what are the units of the rate constant, k?

(b) For each of the following changes, deduce the effect on the rate of reaction:

(i) the concentration of A is increased threefold

(ii) the concentration of B is halved

(iii) the concentrations of A, B and C are all doubled

Answers on p. 102

Determination of a two-step reaction mechanism

Revised

The rate-determining step is defined as the slowest step in the reaction.

The rate equation can provide clues about a likely reaction mechanism by identifying the slowest stage of a reaction sequence. For instance, if the rate equation is:

$$\text{rate} = k[\text{A}]^2[\text{B}]$$

the slow step will involve two molecules of A and one molecule of B. If the rate equation is:

$$\text{rate} = k[\text{A}][\text{B}]^2$$

the slow step will involve one molecule of A and two molecules of B.

The orders in the rate equation match the number of species involved in the rate-determining step.

Reaction mechanisms often involve many separate steps. You may be asked to use the rate equation and the balanced equation to predict a mechanism for a reaction. In a two-step mechanism, the rate equation indicates the number of molecules of each reactant involved in the **slow** step. The slow step plus the fast step gives the balanced equation.

Example

$$2H_2(g) + 2NO(g) \rightarrow 2H_2O(l) + N_2(g)$$

The rate equation is:

$$\text{rate} = k[H_2(g)][NO(g)]^2.$$

Predict a two-step mechanism.

Answer

The rate equation tells us that this involves one molecule of $H_2(g)$ and two molecules of $NO(g)$.

A possible two-step mechanism is:

Slow step: $1H_2(g) + 2NO(g) \rightarrow H_2O(l) + N_2(g) + O(g)$

+

Fast step: $1H_2(g) + O(g) \rightarrow H_2O(l)$

Balanced equation: $2H_2(g) + 2NO(g) \rightarrow 2H_2O(l) + N_2(g)$

How far?

The section on chemical equilibrium in Unit F322: Chains, Energy and Resources introduced the idea of a **reversible reaction**, a **dynamic equilibrium** and **le Chatelier's principle**.

In a **reversible reaction** the reagents react to form products and the products react to re-form the reagents. The reaction proceeds in both forward and reverse directions, which leads to the formation of a dynamic equilibrium.

A **dynamic equilibrium** is reached when the rate of the forward reaction equals the rate of the reverse reaction. The concentrations of the reagents and products remain constant; the reagent and the product molecules react continuously.

Le Chatelier's principle states that if a closed system at equilibrium is subject to a change, the system will move to *minimise* the effect of that change.

Effect of change on a system at equilibrium

Revised

Temperature

The effect of changing temperature depends on whether the forward reaction is exothermic or endothermic:

- If the forward reaction is exothermic, $-\Delta H$, increasing the temperature moves the equilibrium position to the left.
- If the forward reaction is endothermic, $+\Delta H$, increasing the temperature moves the equilibrium position to the right.

Pressure

The effect of changing pressure depends on the number of molecules of gas on each side of the equilibrium:

- Increasing the pressure moves the equilibrium position to the side with fewer molecules of gas.
- Decreasing the pressure moves the equilibrium position to the side with most molecules of gas.

Concentration

The effect of increasing the concentration of either a reagent or a product is to move the equilibrium position in the opposite direction.

Catalysts

Catalysts speed up the rate of the reaction but do *not* change the position of the equilibrium.

You should be able to use le Chatelier's principle to deduce what happens to the position of equilibrium when the system is subjected to change. For example:

$$2NO_2(g) \rightleftharpoons N_2O_4(g) \qquad \Delta H = +100\,\text{kJ mol}^{-1}$$

Increase in temperature	Increase in pressure	Use of a catalyst	Increase in concentration of NO_2
Equilibrium moves to the right because the forward reaction is endothermic	Equilibrium moves to the right because there are fewer molecules of gas on the right-hand side of the equilibrium	Equilibrium position remains unchanged because a catalyst speeds up the forward and reverse reactions equally	Equilibrium moves to the right to reduce the amount of NO_2 in the equilibrium

The equilibrium constant, K_c

Revised

The equilibrium law

K_c is the equilibrium constant in terms of equilibrium concentrations. The equilibrium law states that, for an equation:

$$aA + bB \rightleftharpoons cC + dD$$

$$K_c = \frac{[C]^c[D]^d}{[A]^a[B]^b}$$

[A], [B], [C] and [D] are the **equilibrium concentrations** of the reactants and products in the reaction.

Each product and reactant has its equilibrium concentration raised to the **power** of its **balancing number** (*a*, *b*, *c* and *d*) in the equation.

Working out K_c

Revised

Consider the equilibrium:

$$H_2(g) + I_2(g) \rightleftharpoons 2HI(g)$$

Applying the equilibrium law:

$$K_c = \frac{[HI(g)]^2}{[H_2(g)][I_2(g)]}$$

At equilibrium: $[H_2(g)] = 0.012\ mol\ dm^{-3}$; $[I_2(g)] = 0.001\ mol\ dm^{-3}$; $[HI(g)] = 0.025\ mol\ dm^{-3}$.

$$K_c = \frac{(0.025)^2}{0.012 \times 0.001} = 52.1$$

Units of K_c

The units of K_c depend upon the equilibrium expression for the reaction. Each concentration value is replaced by its units:

$$K_c = \frac{[HI(g)]^2}{[H_2(g)][I_2(g)]} = \frac{(mol\ dm^{-3})^2}{(mol\ dm^{-3})(mol\ dm^{-3})}$$

In this equilibrium, the units cancel and K_c has no units.

Writing expressions for K_c

It is essential that you are able to write expressions for K_c and are able to deduce the units, if any, for each expression. Some examples are given in the table below.

Equilibrium	$2NO_2(g) \rightleftharpoons N_2O_4(g)$	$N_2(g) + 3H_2(g) \rightleftharpoons 2NH_3(g)$	$Br_2(g) + H_2(g) \rightleftharpoons 2HBr(g)$
K_c	$K_c = \frac{[N_2O_4(g)]}{[NO_2(g)]^2}$	$K_c = \frac{[NH_3(g)]^2}{[N_2(g)][H_2(g)]^3}$	$K_c = \frac{[HBr(g)]^2}{[H_2(g)][Br_2(g)]}$
Units	$mol^{-1}\ dm^3$	$mol^{-2}\ dm^6$	None

Properties of K_c

Revised

K_c indicates how *far* a reaction proceeds but tells us nothing about how *fast* the reaction occurs. The size of K_c indicates the extent of a chemical equilibrium:

- If K_c is big ($K_c = 1000$), the equilibrium lies to the right-hand side and a high percentage of product is formed.
- If K_c is small ($K_c = 1 \times 10^{-3}$), the equilibrium lies to the left-hand side and a low percentage of product is formed.
- If $K_c = 1$, the equilibrium lies halfway between reactants and products.

Changing K_c

Revised

K_c is a constant, *but* it is temperature dependent — it can be changed by altering the temperature. K_c is unaffected by changes in concentration or pressure.

- In an exothermic reaction, K_c decreases with increasing temperature because raising the temperature reduces the equilibrium yield of products.

- In an endothermic reaction, K_c increases with increasing temperature because raising the temperature increases the equilibrium yield of products.

Determination of K_c by experiment

The equilibrium constant, K_c can be determined from experimental results. The example illustrates how to answer a typical question.

Example

0.200 mol CH_3COOH and 0.100 mol C_2H_5OH were mixed together with a trace of acid catalyst in a total volume of 200 cm^3. The mixture was allowed to reach equilibrium:

$CH_3COOH + C_2H_5OH \rightleftharpoons CH_3COOC_2H_5 + H_2O$

Analysis of the mixture showed that 0.115 mol of CH_3COOH were present at equilibrium. Calculate the equilibrium constant, K_c.

Answer

Step 1: Find the change in moles of each component in the equilibrium.

From the information given, the number of moles of CH_3COOH that reacted = 0.200 − 0.115 = 0.085

The balanced equation tells us the molar ratio of the reactants and the products.

Balanced equation:	CH_3COOH +	$C_2H_5OH \rightleftharpoons$	$CH_3COOC_2H_5$ +	H_2O
Molar ratio:	1 mol	1 mol	1 mol	1 mol
Change in moles:	−0.085	−0.085	+0.085	+0.085

Step 2: Determine the equilibrium concentration of each component.

	CH_3COOH +	C_2H_5OH	$\rightleftharpoons CH_3COOC_2H_5$ +	H_2O
Initial amount/mol	0.200	0.100	0	0
Change in moles	−0.085	−0.085	+0.085	+0.085
Equilibrium amount/mol	0.115	0.015	0.085	0.085
Equilibrium concentration/mol dm^{-3}	0.115/0.20	0.015/0.20	0.085/0.20	0.085/0.20

Step 3: Write the expression for K_c, substitute values and calculate K_c:

$$K_c = \frac{[CH_3COOC_2H_5][H_2O]}{[CH_3COOH][C_2H_5OH]}$$

$$= \frac{0.085/0.20 \times 0.085/0.20}{0.115/0.20 \times 0.015/0.20} = \frac{0.425 \times 0.425}{0.575 \times 0.075} = 4.19$$

K_c has no units because the equilibrium concentration units cancel.

Examiner's tip

When calculating a value for K_c, first check the units. If the number of species on each side of the equilibrium equation is the same then K_c has no units and it doesn't matter whether you use moles or concentrations of each chemical.

Typical mistake

K_c relates to concentration and its value can only be calculated using concentrations. In answer to a question such as: '5 mol of $NO_2(g)$ in a 2 dm^3 flask was allowed to reach the equilibrium $2NO_2(g) \rightleftharpoons N_2O_4(g)$. At equilibrium, the mixture contained 3 mols of $NO_2(g)$. Calculate K_c.', many candidates will simply put the number of moles into the expression for K_c. However, it is essential to first convert the moles into concentrations.

Now test yourself

2 Give the expression for the equilibrium constant, K_c for each of the following reactions. In each case, state the units, if any.

(a) $N_2O_4(g) \rightleftharpoons 2NO_2(g)$

(b) $2N_2O(g) \rightleftharpoons 2N_2(g) + O_2(g)$

(c) $2C(s) + O_2(g) \rightleftharpoons 2CO(g)$

(d) $2CO(g) + O_2(g) \rightleftharpoons 2CO_2(g)$

Answers on p. 102

Acids, bases and buffers

You should be able to define an acid as a proton donor and be able to write equations for the reactions of acids.

Brønsted–Lowry acids and bases

Revised

An acid–base reaction involves **proton transfer**, for example:

$$NaOH(aq) + HCl(aq) \rightarrow NaCl(aq) + H_2O(l)$$

The ionic equation is:

$$H^+(aq) + OH^-(aq) \rightarrow H_2O(l)$$

Acids also react with carbonates:

$$Na_2CO_3(aq) + 2HCl(aq) \rightarrow 2NaCl(aq) + CO_2(g) + H_2O(l)$$

The ionic equation is:

$$CO_3^{2-}(aq) + 2H^+(aq) \rightarrow CO_2(g) + H_2O(l)$$

A **Brønsted–Lowry acid** is a proton donor.

A **Brønsted–Lowry base** is a proton acceptor.

Acid–base pairs

Acids and bases are linked by H^+ as **conjugate acid–base pairs**.

By mixing an acid with a base, an equilibrium is set up between two acid–base conjugate pairs.

In a **conjugate acid–base pair**, the conjugate acid donates H^+ and the conjugate base accepts H^+.

In the forward reaction: →

- **$CH_3COOH(aq)$ donates a H^+ to the water and, therefore, behaves as an acid**
- **H_2O accepts a H^+ from $CH_3COOH(aq)$ and, therefore, behaves as a base**

$$CH_3COOH(aq) + H_2O(l) \rightleftharpoons H_3O^+ + CH_3COO^-(aq)$$

← **In the reverse reaction:**

- **H_3O^+ donates a H^+ to the $CH_3COO^-(aq)$ and, therefore, behaves as an acid**
- **$CH_3COO^-(aq)$ accepts a H^+ from H_3O^+ and, therefore, behaves as a base**

$CH_3COOH(aq)$ and $CH_3COO^-(aq)$ form an acid–base conjugate pair and H_3O^+ and $H_2O(l)$ form a second acid–base conjugate pair.

Consider the equilibrium: $NH_3(g) + H_2O(l) \rightleftharpoons NH_4^+(aq) + OH^-(aq)$

In the forward reaction, the water donates a proton to the ammonia and, therefore, behaves as an acid. The ammonia accepts a proton and, therefore, behaves as a base. In the reverse reaction the ammonium ion is the acid and the hydroxide is the base.

In summary: $NH_3(g) + H_2O(l) \rightleftharpoons NH_4^+(aq) + OH^-(aq)$

Base 2 — Acid 1 — Acid 2 — Base 1

Examiner's tip

Many questions involve one strong acid and one weak acid, to illustrate acid–base conjugate pairs. Remember: the stronger acid (the acid with the lower pH or pK_a value) will donate a proton to the weaker acid.

Typical mistake

'Ethanoic acid is mixed with nitric acid forming an equilibrium containing acid–base conjugate pairs. Complete the equation below by filling in the blanks:

$CH_3CO_2H + HNO_3 \rightleftharpoons ... + ...$'

Both $CH_3CO_2H + HNO_3$ are acids and because acids donate protons, the most common *incorrect* answer is $CH_3CO_2^-$ and NO_3^-. The correct response is $CH_3CO_2H_2^+ + NO_3^-$.

Now test yourself

3 For each of the following equilibria, identify the conjugate acid–base pairs.

(a) $HCO_3^- + H_2O \rightleftharpoons H_2CO_3 + OH^-$

(b) $HCO_3^- + OH^- \rightleftharpoons H_2O + CO_3^{2-}$

Answers on p. 102

Strength of acids and bases

Revised

The acid–base equilibrium of an acid, HA, in water is:

$$HA(aq) + H_2O(l) \rightleftharpoons H_3O^+(aq) + A^-(aq)$$

To emphasise the loss of a proton, H^+, by dissociation, the equilibrium can be expressed as:

$$HA(aq) \rightleftharpoons H^+(aq) + A^-(aq)$$

The strength of an acid shows the extent of dissociation into H^+ and A^-.

Strong acids

Acids vary considerably in the ease with which they are able to release their hydrogen ions. Strong acids, such as nitric acid HNO_3, are good proton donors.

equilibrium →

$$HNO_3(aq) \rightleftharpoons H^+(aq) + NO_3^-(aq)$$

The equilibrium position is well to the right. There is almost complete dissociation and it is usual to write the equation as:

$$HNO_3(aq) \rightarrow H^+(aq) + NO_3^-(aq)$$

Weak acids

Weak acids, such as ethanoic acid, CH_3COOH, are poor proton donors. The equilibrium position is well to the left.

← equilibrium

$$CH_3COOH(aq) \rightleftharpoons H^+(aq) + CH_3COO^-(aq)$$

There is only partial dissociation.

Strength and concentration

You must be able to distinguish between the terms 'strong' and 'concentrated', 'weak' and 'dilute'.

- A **strong acid** is highly ionised in aqueous solution.
- A **concentrated acid** is made by dissolving a large amount of the acid in a small volume of water.
- A **weak acid** is only partially ionised in aqueous solution.
- A **dilute acid** is made by dissolving a small amount of the acid in a large volume of water.

The usual way of indicating the strength of an acid is to use the equilibrium constant, K_a, for its ionisation in water.

The acid dissociation constant, K_a

The extent of acid dissociation is shown by an equilibrium constant called the **acid dissociation constant**, K_a.

$$HA(aq) \rightleftharpoons H^+(aq) + A^-(aq)$$

$$K_a = \frac{[H^+(aq)][A^-(aq)]}{[HA(aq)]}$$

$$\text{Units: } K_a = \frac{(\text{mol dm}^{-3})^2}{(\text{mol dm}^{-3})} = \text{mol dm}^{-3}$$

A **high K_a** value shows that the extent of dissociation is large — the acid is strong.

A **low K_a** value shows that the extent of dissociation is small — the acid is weak.

Ethanoic acid ionises as:

$$CH_3COOH(aq) \rightleftharpoons CH_3COO^-(aq) + H^+(aq)$$

$$K_a = \frac{[CH_3COO^-(aq)][H^+(aq)]}{[CH_3COOH(aq)]}$$

The K_a value for ethanoic acid is 1.7×10^{-5} mol dm^{-3}. This very low value indicates that a solution of ethanoic acid consists largely of ethanoic acid molecules with relatively few ethanoate ions and hydrogen ions.

The K_a value for methanoic acid is 1.6×10^{-4} mol dm^{-3}, which is almost ten times larger than the value for ethanoic acid. This indicates that methanoic acid, though weak, is stronger than ethanoic acid.

The mineral acids have much higher values for K_a. The value for nitric acid is approximately 40 mol dm^{-3}; that for sulfuric acid is often just listed as 'very large'.

Calculating hydrogen ion concentrations

Revised

The pH scale

The concentration of $H^+(aq)$ ions in acid solutions varies widely between about 10 mol dm^{-3} and about 1×10^{-15} mol dm^{-3}. The **pH** scale is used to overcome the problem of this wide range of numbers. It is a logarithmic scale — each change of one unit on the pH scale corresponds to a tenfold change in the $H^+(aq)$ concentration.

pH is defined by the equation:
$pH = -\log_{10}[H^+(aq)]$

pH	0	1	2	3	4	5	6	7	8	9	10	11	12	13	14
$[H^+]$	1	10^{-1}	10^{-2}	10^{-3}	10^{-4}	10^{-5}	10^{-6}	10^{-7}	10^{-8}	10^{-9}	10^{-10}	10^{-11}	10^{-12}	10^{-13}	10^{-14}

More acidic ← Neutral → More alkaline

You should be able to convert pH to $H^+(aq)$ and vice versa using the relationships shown below.

$$pH = -\log_{10}[H^+(aq)] \text{ and } [H^+(aq)] = 10^{-pH}$$

Calculating the pH of strong acids

For a strong acid, we can assume **complete dissociation** and the concentration of $H^+(aq)$ can be found from the acid concentration.

Example 1

A strong acid, HA, has a concentration of 0.020 mol dm^{-3}. What is the pH?

Answer

The acid dissociates completely. Therefore:

$[H^+(aq)] = 0.020$ mol dm^{-3}

$pH = -\log_{10}[H^+(aq)] = -\log_{10}(0.020) = 1.7$

Example 2

A strong acid, HA, has a pH of 2.4. What is the concentration of $H^+(aq)$?

Answer

The acid dissociates completely. Therefore:

$[H^+(aq)] = 10^{-pH} = 10^{-2.4}$ mol dm^{-3}

$[H^+(aq)] = 3.98 \times 10^{-3}$ mol dm^{-3}

Calculating the pH of weak acids

Weak acids do not dissociate completely. To calculate the pH of a weak acid, HA, you need to know:

- the concentration of the acid
- the acid dissociation constant, K_a

In the equilibrium of a weak aqueous acid, HA(aq), we assume that only a very small proportion of HA dissociates. Hence the amount of undissociated acid is taken to be the same as the initial concentration of the acid.

We also assume that $[H^+(aq)]$ equals $[A^-(aq)]$.

Using these approximations:

$$K_a = \frac{[H^+(aq)][A^-(aq)]}{[HA(aq)]} = \frac{[H^+(aq)]^2}{[HA(aq)]}$$

Example

For a weak acid, $[HA(aq)] = 0.200\ mol\ dm^{-3}$; $K_a = 1.70 \times 10^{-4}\ mol\ dm^{-3}$ at 25°C.

Calculate the pH.

Answer

$$K_a = \frac{[H^+(aq)][A^-(aq)]}{[HA(aq)]} = \frac{[H^+(aq)]^2}{[[HA(aq)]}$$

Therefore:

$$1.70 \times 10^{-4} = \frac{[H^+(aq)]^2}{0.200}$$

Rearranging gives:

$$(1.70 \times 10^{-4}) \times 0.200 = [H^+(aq)]^2$$

$$[H^+(aq)] = \sqrt{(1.70 \times 10^{-4}) \times 0.200} = 5.83 \times 10^{-3} = 0.00583$$

$$pH = -\log_{10}[H^+(aq)] = -\log_{10}(0.00583) = 2.23$$

An alternative way of doing this type of calculation is to use the equation:

$$pH = -\log_{10}\sqrt{K_a \times [HA(aq)]}$$

4 Calculate the pH of each of the following aqueous solutions:

(a) $0.15\ mol\ dm^{-3}\ HNO_3$

(b) $0.15\ mol\ dm^{-3}$ HCN ($K_a = 4.8 \times 10^{-10}\ mol\ dm^{-3}$)

(c) $0.15\ mol\ dm^{-3}$ NaOH

5 Calculate the pH of a mixture of $20.0\ cm^3$ of $1.00\ mol\ dm^{-3}$ HCl and $10.0\ cm^3$ of $1.00\ mol\ dm^{-3}$ NaOH.

Answers on p. 102

K_a and pK_a

Revised

As with $[H^+ (aq)]$ and pH, K_a is often expressed as the logarithmic form, $\mathbf{pK_a}$, which is defined as:

$$pK_a = -\log_{10}K_a.$$

It is a more convenient way of comparing acid strengths.

On this logarithmic scale, each change of one unit on the pK_a scale corresponds to a tenfold change in K_a. Like pH, pK_a can be used as a guide to acidity. The lower the pK_a value, the stronger the acid.

Acid		K_a/mol dm^{-3}	pK_a
Ethanoic acid	CH_3COOH	1.7×10^{-5}	$-\log_{10}(1.7 \times 10^{-5}) = 4.8$
Benzoic acid	C_6H_5COOH	6.3×10^{-5}	$-\log_{10}(6.3 \times 10^{-5}) = 4.2$

This indicates that benzoic acid is a stronger acid than ethanoic acid.

The ionic product of water, K_w

Revised

Water ionises very slightly, acting as both an acid and a base:

$H_2O(l)$ +	$H_2O(l)$ ⇌	$H_3O^+(aq)$ +	$OH^-(aq)$
Acid 1	Base 2	Acid 2	Base 1
(donates proton)	(accepts proton)	(donates proton)	(accepts proton)

More simply:

$$H_2O(l) \rightleftharpoons H^+(aq) + OH^-(aq).$$

In water, only a very small proportion of molecules dissociates into $H^+(aq)$ and $OH^-(aq)$ ions and the equilibrium lies well to the left.

Treating water as a weak acid:

$$K_a = \frac{[H^+(aq)][OH^-(aq)]}{[H_2O(l)]}$$

Rearranging gives:

$$K_a \times [H_2O(l)] = [H^+(aq)][OH^-(aq)]$$

- $K_a \times [H_2O(l)]$ is a constant, K_w, and is called the **ionic product** of water.
- $K_w = [H^+(aq)][OH^-(aq)] = 1.0 \times 10^{-14}\,mol^2\,dm^{-6}$ (at 25°C)
- K_w is temperature dependent and is equal to $1.0 \times 10^{-14}\,mol^2\,dm^{-6}$ at 25°C (298 K) only. (At 10°C, (283 K), $K_w = 2.9 \times 10^{-15}\,mol^2\,dm^{-6}$; at 40°C (313 K), $K_w = 2.9 \times 10^{-14}\,mol^2\,dm^{-6}$).

K_w can be used to calculate the pH of water

At 25°C: $K_w = [H^+(aq)][OH^-(aq)] = 1.0 \times 10^{-14}\,mol^2\,dm^{-6}$

Assuming that $[H^+(aq)] = [OH^-(aq)]$, then $K_w = [H^+(aq)]^2 = 1.0 \times 10^{-14}\,mol^2\,dm^{-6}$

$[H^+(aq)] = 1.0 \times 10^{-7}\,mol\,dm^{-3}$

$pH = -\log_{10}[H^+(aq)] = -\log_{10}(1.0 \times 10^{-7})$

$pH = 7.0$

However, since K_w changes with temperature it follows that the pH of water is equal to 7.0 at 25°C only.

At 10°C:

$K_w = [H^+(aq)][OH^-(aq)] = [H^+(aq)]^2$

$K_w = [H^+(aq)]^2 = 2.9 \times 10^{-15}\,mol^2\,dm^{-6}$

$[H^+(aq)] = 5.4 \times 10^{-8}\,mol\,dm^{-3}$

$pH = -\log_{10}[H^+(aq)] = -\log_{10}(5.4 \times 10^{-8})$

$pH = 7.3$

At 40°C:

$K_w = [H^+(aq)][OH^-(aq)] = [H^+(aq)]^2$

$K_w = [H^+(aq)]^2 = 2.9 \times 10^{-14}\,mol^2\,dm^{-6}$

$[H^+(aq)] = 1.7 \times 10^{-7}\,mol\,dm^{-3}$

$pH = -\log_{10}[H^+(aq)] = -\log_{10}(1.7 \times 10^{-7})$

$pH = 6.8$

The pH of water can be calculated using $pH = -\log\sqrt{K_w}$

K_w can be used to calculate the pH of strong alkalis

The pH of a strong alkali, such as NaOH, can be calculated from the concentration of the alkali and the ionic product of water, $\boldsymbol{K_w}$.

Example

A strong alkali, KOH, has a concentration of 0.50 mol dm^{-3}. What is the pH at 25°C?

Answer

KOH dissociates completely:

$$KOH(aq) \rightarrow K^+(aq) + OH^-(aq)$$

Therefore:

$$[OH^-(aq)] = [KOH(aq)] = 0.50\ mol\ dm^{-3}$$

$$K_w = [H^+(aq)][OH^-(aq)] = 1 \times 10^{-14}\ mol^2\ dm^{-6}$$

$$[H^+(aq)] = \frac{K_w}{[OH^-(aq)]} = \frac{1 \times 10^{-14}}{0.50} = 2 \times 10^{-14}\ mol\ dm^{-3}$$

$$pH = -\log_{10}[H^+(aq)]$$

$$= -\log_{10}(2 \times 10^{-14}) = 13.7$$

Buffer solutions

Revised

A buffer solution resists changes in pH during the addition of an acid or an alkali. The buffer solution maintains a near constant pH by removing most of any acid or alkali that is added to the solution.

A buffer solution can be a mixture of a weak acid, HA, and its conjugate base, A^-:

$$\underset{\text{Weak acid}}{HA(aq)} \rightleftharpoons H^+(aq) + \underset{\text{Conjugate base}}{A^-(aq)}$$

An example of a common buffer solution is a mixture of CH_3COOH (the weak acid) and $CH_3COO^-Na^+$ (the conjugate base). The pH at which the buffer operates depends on the K_a of the weak acid and the relative concentrations of the weak acid and the conjugate base.

In a mixture of CH_3COOH and $CH_3COO^-Na^+$, the CH_3COOH partially dissociates giving low concentrations of $CH_3COO^-(aq)$ and $H^+(aq)$:

$$CH_3COOH(aq) \rightleftharpoons CH_3COO^-(aq) + H^+(aq)$$

$CH_3COO^-Na^+$ dissociates completely giving high concentrations of $CH_3COO^-(aq)$:

$$CH_3COO^-Na^+(aq) \rightarrow CH_3COO^-(aq) + Na^+(aq)$$

The high $CH_3COO^-(aq)$ concentration forces the equilibrium to the left-hand side and results in the buffer solution containing a low concentration of $H^+(aq)$ and high concentrations of $CH_3COOH(aq)$ and $CH_3COO^-(aq)$.

A buffer solution contains *two* important components:

- a high concentration of the weak acid, for example $CH_3COOH(aq)$
- a high concentration of the conjugate base, for example $CH_3COO^-(aq)$

On addition of an acid, $H^+(aq)$, the high concentration of the conjugate base $CH_3COO^-(aq)$ removes most of the added $H^+(aq)$ by forming $CH_3COOH(aq)$:

$$CH_3COO^-(aq) + H^+(aq) \rightarrow CH_3COOH(aq)$$

On addition of an alkali, $OH^-(aq)$, the high concentration of $CH_3COOH(aq)$ removes most of the added $OH^-(aq)$ by forming $CH_3COO^-(aq)$:

$$CH_3COOH\,(aq) + OH^-(aq) \rightarrow H_2O(l) + CH_3COO^-(aq)$$

A buffer cannot cancel out the effect of any acid or alkali that is added. The buffer removes most of the added acid or alkali and *minimises* any changes in pH.

Calculations involving buffer solutions

The pH of a buffer solution depends upon the acid dissociation constant, K_a, of the acid and the molar ratio of the weak acid and its conjugate base.

For a buffer containing the weak acid, HA, and its conjugate base, A^-:

$$K_a = \frac{[H^+(aq)][A^-(aq)]}{[HA(aq)]}$$

Therefore:

$$[H^+(aq)] = K_a \frac{[HA(aq)]}{[A^-(aq)]}$$

$pH = -\log[H^+(aq)]$, so

$$pH = -\log \frac{K_a[HA(aq)]}{[A^-(aq)]}$$

It follows that the pH of a buffer can be altered by adjusting the weak acid to conjugate base ratio.

An alternative approach is to use the Henderson–Hasselbach equation which states:

$$pH = pK_a + \log\frac{[A^-]}{[HA]}$$

This can be used to calculate the concentrations of the weak acid and of the salt of the weak acid that must be used to produce a buffer at a fixed pH. The Henderson–Hasselbach equation can be rearranged to give:

$$\frac{[A^-]}{[HA]} = 10^{-(pH-pK_a)}$$

Example

(a) Calculate the pH of a buffer with concentrations of $0.10\,mol\,dm^{-3}$ $CH_3COOH(aq)$ and $0.10\,mol\,dm^{-3}$ $CH_3COO^-(aq)$. For CH_3COOH, $K_a = 1.7 \times 10^{-5}\,mol\,dm^{-3}$.

(b) What happens to the pH if the concentration of $CH_3COOH(aq)$ is changed to $0.30\,mol\,dm^{-3}$?

Answer

(a) First, calculate $[H^+(aq)]$:

$$[H^+(aq)] = \frac{K_a[HA(aq)]}{[A^-(aq)]}$$

$$[H^+(aq)] = \frac{1.7 \times 10^{-5} \times 0.10}{0.10} = 1.7 \times 10^{-5}\,mol\,dm^{-3}$$

Then use $[H^+(aq)]$ to calculate pH:

$$pH = -\log_{10} [H^+(aq)]$$
$$= -\log_{10}(1.7 \times 10^{-5}) = 4.77$$

pH of the buffer solution is 4.77.

(b) First, calculate $[H^+(aq)]$

$$[H^+(aq)] = \frac{K_a[HA(aq)]}{[A^-(aq)]}$$

$$[H^+(aq)] = \frac{1.7 \times 10^{-5} \times 0.30}{0.10} = 5.1 \times 10^{-5}\,mol\,dm^{-3}$$

Then use $[H^+(aq)]$ to calculate pH:

$$pH = -\log_{10}[H^+(aq)]$$
$$= -\log_{10}(5.1 \times 10^{-5}) = 4.29$$

pH of the buffer solution is 4.29.

Examiner's tip

Buffers can be prepared from a weak acid and a strong base, providing that the weak acid is in excess. If 100 cm^3 of 0.2 mol dm^{-3} $CH_3COOH(aq)$ is mixed with 100 cm^3 of 0.1 mol dm^{-3} $NaOH(aq)$, the resultant solution contains undissociated $CH_3COOH(aq)$ and $CH_3COO^-(aq)$. This solution is a buffer.

Control of pH in blood

Revised

In the human body the blood plasma has a normal pH of 7.35–7.45. If the pH falls below 7.0 or rises above 7.8 the results could be fatal. The buffer systems in the blood are extremely effective and protect the fluid from large changes in pH. Blood contains a number of buffering systems, the major one being the carbonic acid–hydrogencarbonate system:

$$H_2O(l) + CO_2(g) \rightleftharpoons H_2CO_3(aq) \rightleftharpoons HCO_3^-(aq) + H^+(aq)$$

H_2CO_3 is carbonic acid; HCO_3^- is hydrogencarbonate

Adding an acid to the system increases the concentration of $H^+(aq)$, driving the equilibrium to the left. This increases the concentration of carbonic acid, H_2CO_3, which in turn is decreased by an increased breathing rate. More carbon dioxide is exhaled resulting in more H_2CO_3 breaking down to replace it. The two equilibria together resist the increase in acidity.

Common uses of buffers

Soaps and detergents are alkaline and irritate the skin and eyes. Shampoo often contains a mixture of citric acid (a weak acid) and sodium citrate (the conjugate base) which acts as a buffer. The ratio of the acid and the conjugate base is adjusted so that the pH is maintained at about pH 5.5, which is approximately the pH of skin.

Babies often suffer from nappy rash because dirty nappies contain ammonia, which is a weak base and has a pH in the region 7–9. Bacteria multiply rapidly in the pH region 7–9 but not at pH 6. Baby lotions are buffered around pH 5.5–6.0, the approximate pH of skin, so that the lotion protects the baby by preventing bacteria from multiplying.

pH changes and indicators

Revised

Indicators are substances that change colour with a change in pH. Many indicators are weak acids and can be represented by HIn. The weak acid, HIn, and its conjugate base, In^-, have different colours, for example for methyl orange:

RED				YELLOW
HIn(aq)	$\rightleftharpoons$	$H^+(aq)$	+	$In^-(aq)$
Weak acid				Conjugate base

At the end point of a titration HIn (red) and In^- (yellow) are present in equal concentrations. Therefore, the colour at the end point is orange. The pH at the end point is equal to the pK_{in} of the indicator.

pH ranges for common indicators

An indicator changes colour over a range of about two pH units within which is the pK_{in} value of the indicator.

pH	0	1	2	3	4	5	6	7	8	9	10	11	12	13	14
			Red ↔ Yellow Methyl orange, pK_{in} = 3.7												
								Colourless ↔ Pink Phenolphthalein, pK_{in} = 9.3							

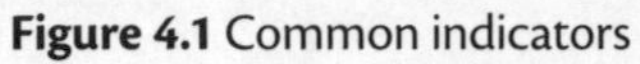

Figure 4.1 Common indicators

Choosing an indicator

When the acid and the base have completely reacted, this is known as the **equivalence point**. At the equivalence point of the titration there is a sharp change in pH for a very small addition of acid or base.

The choice of a suitable indicator is best shown using titration curves, which show the changes in pH during a titration.

Key features of titration curves

- The pH changes rapidly at the near vertical portion of the titration curve. This is the end point of the titration.
- The sharp change in pH is brought about by a very small addition of alkali, typically the addition of one drop.
- The indicator is only suitable if its pK_{in} value is within the pH range of the near vertical portion of the titration curve.

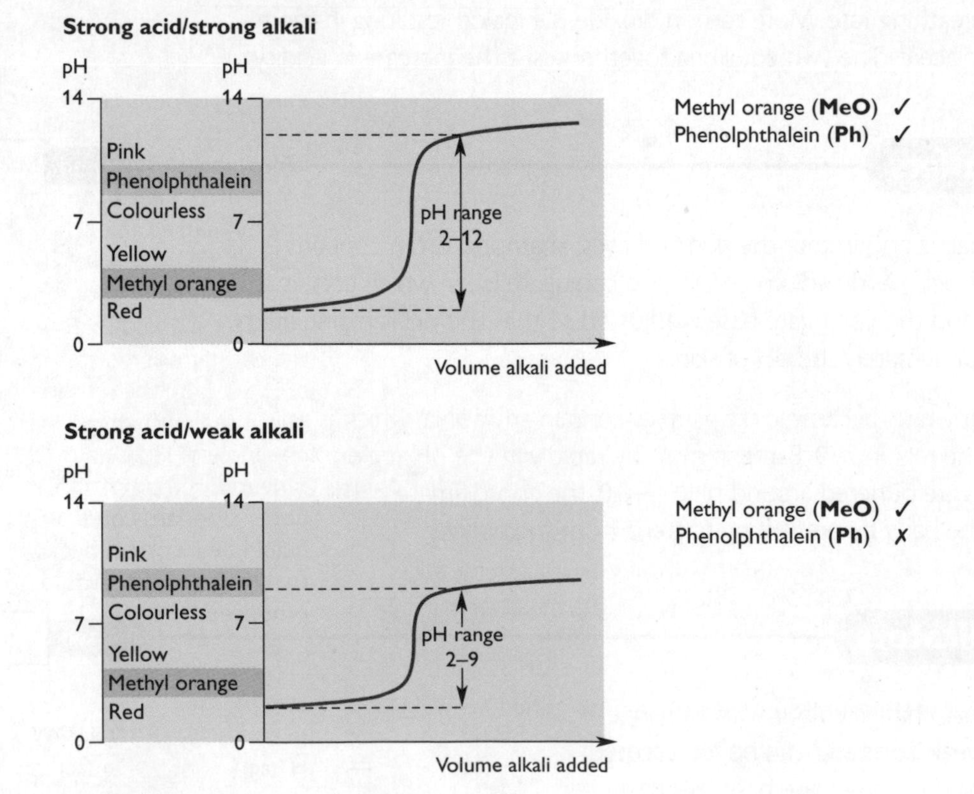

Weak acid/strong alkali

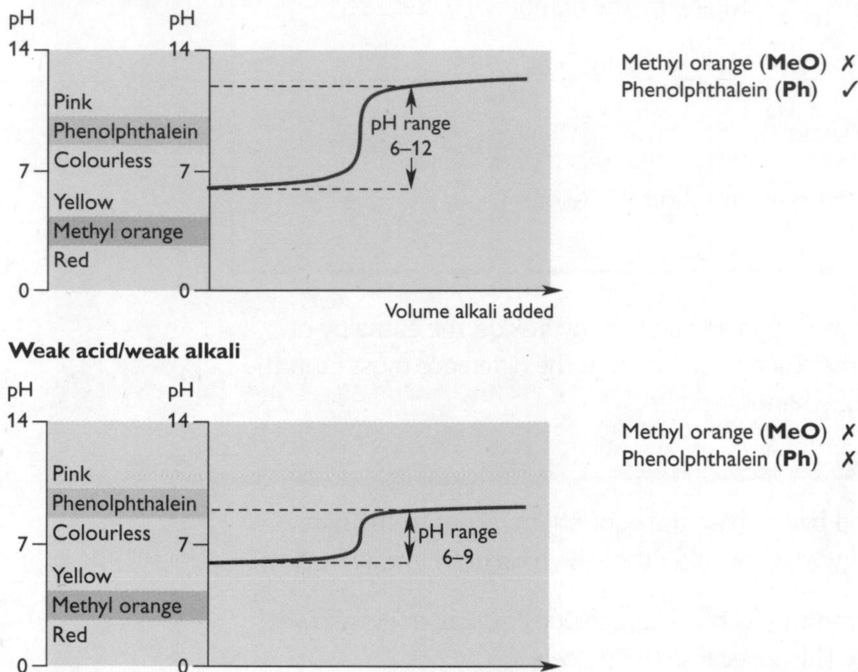

Neutralisation

Enthalpy of neutralisation can be used to distinguish between strong and weak acids.

The **standard enthalpy of neutralisation** is the enthalpy change that occurs when one mole of water is produced in the reaction of an acid with an alkali under standard conditions.

The standard enthalpy of neutralisation is defined in terms of the amount of water formed in order to take into account the basicity of the acid. If the acid is dibasic — for example sulfuric acid — 1 mol of the acid would require twice as much sodium hydroxide to neutralise it compared with 1 mol of monobasic hydrochloric acid:

$$H_2SO_4(aq) + 2NaOH(aq) \rightarrow Na_2SO_4(aq) + 2H_2O(l)$$

$$HCl(aq) + NaOH(aq) \rightarrow NaCl(aq) + H_2O(l)$$

If the enthalpy of neutralisation is quoted for the formation of 1 mol of water, this difference is removed and the figures obtained for the enthalpy changes can be compared directly. In fact, for sulfuric and hydrochloric acids, the enthalpy change per mole of water produced is the same.

This is not really surprising because the reaction taking place is the same:

$$H^+(aq) + OH^-(aq) \rightarrow H_2O(l)$$

Enthalpy of neutralisation can be calculated from experimental data using the equation:

$$\Delta H = \frac{Q}{n} = \frac{mc\Delta T}{n}$$

Typical mistake

If asked to define the enthalpy of neutralisation of HCl(aq) it would be acceptable to state 'the enthalpy change when one mole of HCl(aq) is neutralised by an alkali under standard conditions.' However, if asked to define the enthalpy of neutralisation of $H_2SO_4(aq)$, the response 'the enthalpy change when one mole of $H_2SO_4(aq)$ is neutralised by an alkali under standard conditions' would be incorrect because two moles of H_2O would be produced.

Example

When 50.0 cm^3 of 2.00 mol dm^{-3} HCl was mixed with 50.0 cm^3 of 2.00 mol dm^{-3} NaOH, the temperature increased by 13.7°C. Calculate the standard enthalpy change of neutralisation of HCl(aq). Assume that the specific heat capacity, c, is 4.20 J g^{-1} K^{-1} and that the density of both HCl(aq) and NaOH(aq) is 1.00 g cm^{-3}.

Answer

Step 1: Calculate the energy transferred in the reaction by using $Q = mc\Delta T$

mass of the two solutions, m = 50.0 + 50.0 = 100.0 g

$Q = mc\Delta T$ = 100.0 × 4.20 × 13.7 = 5754 J = 5.754 kJ

Step 2: Convert the answer to $kJ\,mol^{-1}$ by dividing by the number of moles used.

amount n, in moles, of HCl $= cV = 2.00 \times \frac{50.0}{1000} = 0.100$ mol

$\Delta H = \frac{Q}{n} = \frac{5.754}{0.100} = 57.54 = 57.5\ kJ\,mol^{-1}$

The temperature rose, so it is an exothermic reaction. Therefore:

$\Delta H = -57.5\ kJ\,mol^{-1}$

If the experiment is repeated using ethanoic acid and sodium hydroxide, the enthalpy of neutralisation obtained is lower. Since the reaction is the same, the difference must lie in the strength of the acid. The dissociation of ethanoic acid:

$$CH_3COOH(aq) \rightleftharpoons CH_3COO^-(aq) + H^+(aq)$$

is not complete. As the H^+ is neutralised by the base the equilibrium moves to the right. This requires a certain amount of energy which would otherwise have been lost as heat.

The neutralisation reaction of aqueous ammonia and hydrochloric acid has an even lower value for the enthalpy of neutralisation. This is because the process:

$$NH_3(g) + H_2O(l) \rightleftharpoons NH_4^+(aq) + OH^-(aq)$$

requires even more energy than the dissociation of ethanoic acid.

The neutralisation of the weak acid–weak base combination of ethanoic acid and ammonia results in an even lower value for the enthalpy of neutralisation.

Experiments of this type can give some indication of the strengths of acids and bases, even though the measurements are not particularly accurate.

Check your understanding

1 The table below gives data for the reaction:

$$BrO_3^-(aq) + 5Br^-(aq) + 6H^+(aq) \rightarrow 3Br_2(aq) + 3H_2O(l)$$

Experiment	$[BrO_3^-(aq)]/mol\,dm^{-3}$	$[Br^-(aq)]/mol\,dm^{-3}$	$[H^+(aq)]/mol\,dm^{-3}$	Initial rate/$mol\,dm^{-3}\,s^{-1}$
1	0.1	0.2	0.1	1.64×10^{-3}
2	0.2	0.1	0.1	1.64×10^{-3}
3	0.2	0.2	0.1	3.28×10^{-3}
4	0.2	0.1	0.2	6.56×10^{-3}
5	0.25	0.25	0.25	X

(a) Determine the rate equation for this reaction.

(b) Calculate the value of k, including its units.

(c) What would be the initial rate of reaction, x, for the initial concentrations shown in the final row of the table?

2 Nitrogen(I) oxide, N_2O, decomposes to form nitrogen and oxygen according to the equation:

$$2N_2O(g) \rightleftharpoons 2N_2(g) + O_2(g)$$

In an experiment, 1.00 mol of nitrogen(I) oxide is heated in a $1.00\,dm^3$ container until equilibrium is established. The mixture is then analysed and found to contain 0.10 mol of nitrogen(I) oxide.

(a) Calculate the concentrations of nitrogen and oxygen present in the equilibrium mixture.

(b) Calculate the equilibrium constant, K_c.

(c) If the experiment were repeated using 1.00 mol of nitrogen(I) oxide in a $2.00\,dm^3$ container, how would the value of K_c change?

3 (a) The pH of a solution of ethanoic acid is 2.70. K_a for the acid is $1.7 \times 10^{-5}\,mol\,dm^{-3}$. Calculate the concentration of the ethanoic acid solution.

(b) Calculate the mass of sodium ethanoate that must be added to the acid to create a buffer solution of pH 4.0. (Assume that the sodium ethanoate does not cause an increase in volume as it dissolves.)

4 The K_{in} value of chlorophenol red is 6.31×10^{-7}. It is yellow in acid solutions and red in alkaline solutions.

(a) Calculate the pH that is the mid-point for the colour change.

(b) Using chlorophenol red as your example, describe how an indicator works.

(c) Give the colour of chlorophenol red when it is added to the following solutions. Explain your answers.

(i) $0.0001\,mol\,dm^{-3}$ hydrochloric acid

(ii) pure water

5 During the processing of apples the skins can be loosened using sodium hydroxide at pH 12. The pH of the sodium hydroxide eventually drops to 11.5 and it becomes ineffective. This is still very alkaline and so before discarding the solution it is reacted with hydrochloric acid to reduce the pH further to the safer value of 10.8. To ensure this has been done, an indicator called benzaldehyde 3-nitrophenylhydrazone (NPB) is used. NPB is purple in a solution with a pH greater than 12 and yellow at a pH less than 11.

(a) Calculate the change in hydroxide concentration, in $mol\,dm^{-3}$, that takes place as the pH of the solution reduces from 12 to 11.5.

(b) Estimate a value for the equilibrium constant, K_{in}, for the indicator NPB.

(c) Estimate the ratio of the concentration of the un-ionised form of the indicator, NPB, to the concentration of the anion, NPB^-, at a pH of 10.8 (i.e. the ratio $[NPB]/[NPB^-]$).

6 Aspirin is an effective pain killer although its use has, to some extent, been discouraged because in some instances it can cause the stomach to bleed. This appears to be triggered by the molecular form of aspirin dissolving in the covalent lipids in the stomach lining.

Aspirin contains carboxylic acid and ester groups. It is quite readily hydrolysed:

$$C_6H_4(OCOCH_3)CO_2H + H_2O \rightleftharpoons C_6H_4(OH)CO_2H + CH_3COOH \quad K_a = 3 \times 10^{-4}\,mol\,dm^{-3}$$

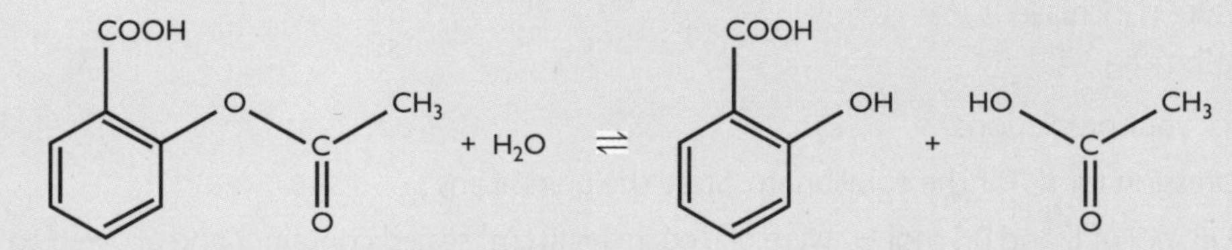

Because of the ease of hydrolysis, aspirin has a limited shelf life.

(a) Assuming that the pH of stomach acid is approximately 1, explain why stomach bleeding might be a problem.

(b) The blood is buffered at a pH of 7.4. Calculate whether aspirin in the blood exists largely in its unionised molecular form or as an anion. (K_a for aspirin is $3 \times 10^{-4}\,mol\,dm^{-3}$.)

Aspirin is usually administered as a calcium salt since this is more soluble. However, as aspirin is hydrolysed rapidly above pH 8.5, care has to be taken in its preparation.

(c) If a solution of calcium hydroxide containing $0.741\,g\,dm^{-3}$ is used to create the calcium salt by a reaction with aspirin, is this likely to result in hydrolysis?

(d) A 0.900 g sample of aspirin becomes damp and absorbs 0.100 g of water. An equilibrium is established and analysis shows that 0.117 g of ethanoic acid is present in the equilibrium mixture. Calculate the percentage of aspirin that has been hydrolysed.

Answers on pp. 102–104

Exam practice

1 The reaction between hydrogen and nitrogen monoxide is a redox reaction and results in the formation of nitrogen and water.

(a) (i) Write a balanced equation for the reaction. [1]

(ii) Identify the oxidising agent in the reaction. Justify your answer. [2]

(b) The rate equation for the reaction is:

rate = $k[H_2(g)][NO(g)]^2$

Using 1.2×10^{-2} mol dm^{-3} $H_2(g)$ and 6.0×10^{-3} mol dm^{-3} NO(g), the initial rate of this reaction was 3.6×10^{-2} mol dm^{-3} s^{-1}. Calculate the rate constant, *k*, for this reaction. Quote your answer to two significant figures. State the units of the rate constant, *k*. [4]

(c) Calculate the initial rate of reaction when each of the following changes is made. Show your working.

(i) The concentration of hydrogen is tripled. [1]

(ii) The concentration of nitrogen monoxide is halved. [1]

(iii) The concentration of both is doubled. [1]

(d) Dinitrogen pentoxide decomposes according to the equation:

$2N_2O_5(g) \rightarrow 4NO_2(g) + O_2(g)$

The decomposition is first order with respect to $N_2O_5(g)$.

The decomposition proceeds by a two-step mechanism with the rate-determining step taking place first.

(i) Write a rate equation for this reaction. [1]

(ii) Explain the term *rate-determining step.* [1]

(iii) Suggest the two steps for this reaction and write their equations. Show clearly that the two steps equate to the balanced equation given above. [3]

2 Hydrogen and iodine react according to the equation:

$H_2(g) + I_2(g) \rightleftharpoons 2HI(g)$ $\Delta H = +53.0$ kJ mol^{-1}.

(a) State le Chatelier's principle. [1]

(b) Use le Chatelier's principle to predict what happens to the position of the equilibrium when:

(i) the temperature is increased

(ii) the pressure is increased

(iii) a catalyst is used

Justify each of your predictions. [6]

(c) Write an expression for K_c for the equilibrium. State the units, if any. [2]

(d) (i) When 0.18 mol of I_2 and 0.5 mol H_2 were placed in a 500 cm^3 sealed container and allowed to reach equilibrium, the equilibrium mixture was found to contain 0.010 mol of I_2. Calculate K_c. [5]

(ii) Explain what would happen to the value of K_c if the experiment were repeated with the 500 cm^3 container replaced by one with a volume of 1 dm^3. [2]

3 (a) (i) A weak organic acid, HA, has the percentage composition by mass: C, 40%; H, 6.7%; O, 53.3%. Calculate the empirical formula of HA. [2]

(ii) The relative molecular mass of HA is 60.0. What is its molecular formula? [1]

(b) 1.20 g of HA were dissolved in 250.0 cm^3 of water. Calculate the pH of the resulting solution. Show all your working.

(K_a of HA = 1.7×10^{-5} mol dm^{-3}) [5]

(c) A 0.04 mol dm^{-3} solution of HA was titrated with a 0.05 mol dm^{-3} solution of sodium hydroxide.

(i) Calculate the pH of the NaOH(aq). ($K_w = 1.0 \times 10^{-14}$ mol^2 dm^{-6}) [2]

(ii) Calculate the volume of NaOH(aq) required to neutralise 25.0 cm^3 of solution HA. [3]

(iii) Sketch a graph to show the change in pH during the titration. [3]

(d) Indicators can be used to determine the end point of a titration. Which of the following would be most suitable for this titration? Justify your answer and suggest what you would see at the end point. [3]

Indicator	Acid colour	pH range	Alkaline colour
Thymol blue (acid)	Red	1.2–2.8	Yellow
Bromocresol purple	Yellow	5.2–6.8	Purple
Thymol blue (base)	Yellow	8.0–9.6	Blue

4 A patient suffering from a duodenal ulcer displays increased acidity in their gastric juices. The exact acidity of the patient's gastric juice is monitored by measuring the pH.

(a) (i) Define pH. **[1]**

(ii) The patient's gastric juice was found to have a hydrochloric acid concentration of 8.0×10^{-2} mol dm^{-3}. Calculate the pH of the gastric juice. **[1]**

(b) One of the most common medications designed for the relief of excess stomach acidity is aluminium hydroxide, $Al(OH)_3$.

(i) Write an equation for the reaction between HCl and $Al(OH)_3$. **[1]**

(ii) On another day, the patient's gastric juice was found to have a pH of 1.3. The patient produces 2 dm^3 of gastric juice in a day. This volume of gastric juice is to be treated with tablets containing $Al(OH)_3$ in order to raise the pH to 2.0. Calculate the mass of aluminium hydroxide required to raise the pH of 2 dm^3 of gastric juice from 1.3 to 2.0. **[5]**

(c) The control of the blood pH is important. This is achieved by the presence of HCO_3^- ions in blood plasma. Using appropriate equations, explain how HCO_3^- acts as a buffer solution. **[3]**

Answers and quick quiz 4 online

Online

Examiner's summary

You should now have an understanding of:

- ✔ orders of reaction and rate equations
- ✔ rate-determining step
- ✔ equilibrium, K_c
- ✔ acids, bases and conjugate pairs
- ✔ pK_a and pH
- ✔ buffers
- ✔ enthalpy of neutralisation

5 Energy

Lattice enthalpy

Review of basic ideas on energetics

Revised

Chemical reactions are usually accompanied by a change in enthalpy (energy), ΔH, normally in the form of heat energy. Reactions tend to be either **exothermic** or **endothermic**:

- An exothermic reaction loses energy to the surroundings and ΔH is negative.
- An endothermic reaction gains energy from the surroundings and ΔH is positive.

Standard enthalpy changes

All standard enthalpy changes are measured under **standard conditions** of temperature and pressure:

- temperature — 298 K (25°C)
- pressure — 101 kPa

Standard conditions are referred to as s.t.p.

The **standard enthalpy change of formation** is the enthalpy change when 1 mol of a substance is formed from its elements, in their natural state, under standard conditions of 298 K and 101 kPa.

1 mol of product must always be formed even if it means using fractions in the balanced equation

$$2C(s) + 3H_2(g) \longrightarrow \mathbf{1}C_2H_6(g)$$

It is essential to show all state symbols

The **standard enthalpy change of combustion** is the enthalpy change when 1 mol of a substance is burnt completely, in an excess of oxygen, under standard conditions of 298 K and 101 kPa.

1 mol of reactant must always be used even if it means using fractions in the balanced equation

$$\mathbf{1}C_2H_6(g) + 3\tfrac{1}{2}O_2(g) \longrightarrow 2CO_2(g) + 3H_2O(l)$$

It is essential to show all state symbols

Average bond enthalpy is the enthalpy change on breaking 1 mol of a covalent bond in a gaseous molecule under standard conditions of 298 K and 101 kPa.

Activation energy

Activation energy is the minimum energy required for colliding particles to react. In any chemical reaction bonds are broken and new bonds are formed. Breaking bonds is an endothermic process that requires energy. The energy requirement contributes to the activation energy of a reaction.

Hess's law

Hess's law states that the enthalpy change for a reaction is the same irrespective of the route taken, provided that the initial and final conditions are the same.

Lattice enthalpy and Born–Haber cycles

Revised

Lattice enthalpy indicates the strength of the ionic bonds in an ionic lattice.

For example:

$$Na^+(g) + Cl^-(g) \rightarrow Na^+Cl^-(s)$$

It is almost impossible to measure lattice enthalpy experimentally, so lattice enthalpy is calculated using a **Born–Haber** cycle. A Born-Haber cycle is similar to a Hess's cycle, enabling calculation of changes that cannot be measured directly.

The **lattice enthalpy ($\Delta H^{\ominus}_{latt}$)** of an ionic compound is the enthalpy change that accompanies the formation of 1 mol of an ionic compound from its constituent gaseous ions. $\Delta H^{\ominus}_{latt.}$ is exothermic.

The lattice enthalpy of sodium chloride can be calculated by considering the standard enthalpy of formation of NaCl(s). In order to form an ionic solid, both sodium and chlorine have to undergo a number of changes. These are outlined in the cycle below.

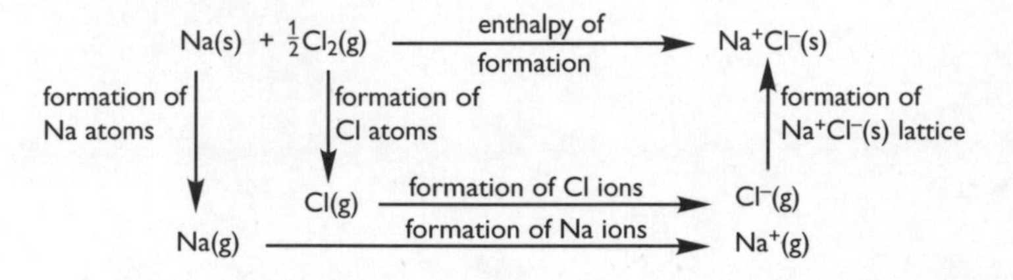

All of the changes in the cycle can be measured experimentally except for the formation of the $Na^+Cl^-(s)$ lattice from its gaseous ions — the lattice enthalpy. However, because enthalpies for the other steps can be measured, the lattice enthalpy can be calculated. The cycle above has to be converted into a Born–Haber cycle, which is a combination of an enthalpy profile diagram and a Hess's cycle. The full Born–Haber cycle for sodium chloride is shown in Figure 5.1

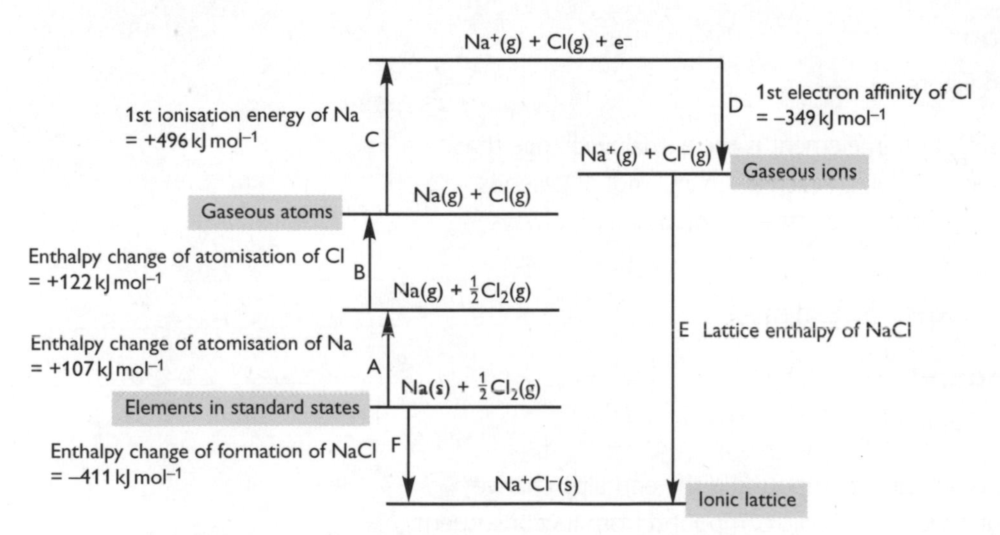

Figure 5.1 Born–Haber cycle for the formation of sodium chloride

Using Hess's law: A + B + C + D + E = F

$$\Delta H^{\ominus}_{at}\,Na(g) + \Delta H^{\ominus}_{at}\,Cl(g) + \Delta H^{\ominus}_{IE}Na(g) + \Delta H^{\ominus}_{EA}Cl(g) + \mathbf{E} = \Delta H^{\ominus}_{f}Na^+Cl^-(s)$$

$$+107 + 122 + 496 + (-349) + \mathbf{E} = -411$$

Hence, the lattice energy of $Na^+Cl^-(s)$, $\mathbf{E} = -787\ kJ\ mol^{-1}$

Examiner's tip

When constructing a Born–Haber cycle it is essential to show the change from *elements* to *gaseous atoms* to *gaseous ions* to *ionic lattice.*

Definitions of enthalpy changes

Revised

Formation of an ionic compound

This is represented by step F in Figure 5.1.

The standard enthalpy change of formation, $\Delta H_f^\ominus$, is the enthalpy change when 1 mol of a substance is formed from its elements, in their natural states, under standard conditions. The standard enthalpy change of formation is usually exothermic for an ionic compound.

$$Na(s) + \frac{1}{2}Cl_2(g) \rightarrow Na^+Cl^-(s) \qquad \Delta H_f^\ominus = -411\ kJ\ mol^{-1}$$

Formation of gaseous atoms

This is represented by steps A and B in Figure 5.1.

The standard enthalpy change of atomisation, $\Delta H_{at}^\ominus$, of an element is the enthalpy change accompanying the formation of 1 mol of gaseous atoms from the element in its standard state. The standard enthalpy change of atomisation is always endothermic.

$$Na(s) \rightarrow Na(g) \qquad \Delta H_{at}^\ominus = +107\ kJ\ mol^{-1}$$

$$\frac{1}{2}Cl_2(g) \rightarrow Cl(g) \qquad \Delta H_{at}^\ominus = +122\ kJ\ mol^{-1}$$

Formation of positive ions

This is represented by step C in Figure 5.1.

The **first ionisation energy ($\Delta H_{IE}^\ominus$)** of an element is the enthalpy change that accompanies the removal of one electron from each atom in 1 mol of gaseous atoms to form 1 mol of gaseous 1+ ions. The first ionisation energy is always endothermic.

$$Na(g) \rightarrow Na+(g) + e^- \qquad \Delta H_{IE}^\ominus = +496\ kJ\ mol^{-1}$$

Formation of negative ions

This is represented by step D in Figure 5.1.

The **first electron affinity ($\Delta H_{EA}^\ominus$)** of an element is the enthalpy change that accompanies the addition of one electron to each atom in 1 mol of gaseous atoms to form 1 mol of gaseous 1– ions. The first electron affinity is always exothermic.

$$Cl(g) + e^- \rightarrow Cl^-(g) \qquad \Delta H_{EA}^\ominus = -349\ kJ\ moll^{-1}$$

Formation of ionic compound

This is represented by step E in Figure 5.1.

The **lattice enthalpy ($\Delta H_{latt}^\ominus$)** of an ionic compound is the enthalpy change accompanying the formation of 1 mol of an ionic compound from its constituent gaseous ions. ($\Delta H_{latt}^\ominus$ is exothermic.)

$$Na^+(g) + Cl^-(g) \rightarrow Na^+Cl^-(s) \qquad \Delta H_{latt}^\ominus = -787\ kJ\ mol^{-1}$$

Calculation of lattice enthalpy

Revised

The lattice enthalpy for magnesium chloride can be calculated using the data shown in Table 5.1.

Table 5.1 Data required for calculating the lattice enthalpies of magnesium chloride

	Standard enthalpy change	Equation	ΔH/kJ mol^{-1}
A	Formation of $MgCl_2(s)$	$Mg(s) + Cl_2(g) \rightarrow MgCl_2(s)$	−641
B	Atomisation of magnesium	$Mg(s) \rightarrow Mg(g)$	+148
C	Atomisation of chlorine	$\frac{1}{2}Cl_2(g) \rightarrow Cl(g)$	+122
D	First ionisation energy of Mg	$Mg(g) \rightarrow Mg^+(g) + 1e^-$	+738
E	Second ionisation energy of Mg	$Mg^+(g) \rightarrow Mg^{2+}(g) + 1e^-$	+1451
F	First electron affinity of Cl	$Cl(g) + 1e^- \rightarrow Cl^-(g)$	−349

The Born–Haber cycle for magnesium chloride is shown in Figure 5.2.

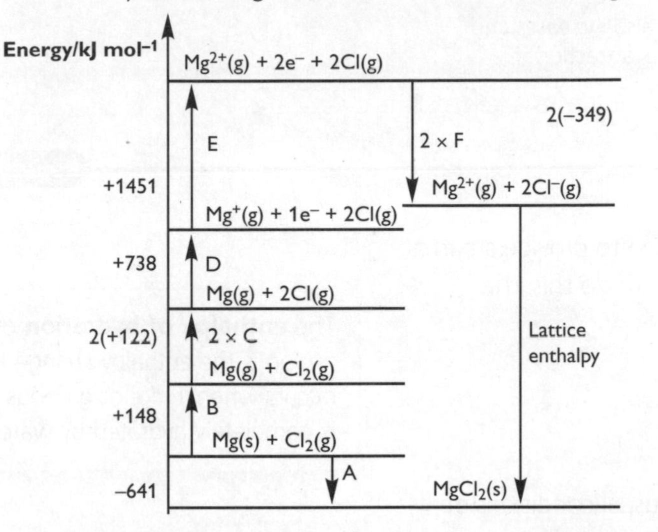

Figure 5.2 Born-Haber cycle for magnesium chloride

Applying Hess's law to the Born–Haber cycle in Figure 5.2 gives:

A = B + 2C + D + E + 2F + lattice enthalphy

Lattice enthalpy = A − B − 2C − D − E − 2F

The lattice enthalpy of magnesium chloride is:

= −641 −148 −244 − 738 −1451 − (−698)

= −2524 kJ mol^{-1}.

Factors affecting the size of lattice enthalpies

Revised

The strength of an ionic lattice and the value of its lattice enthalpy depend upon ionic radius and ionic charge.

Effect of ionic size

Compound	Lattice enthalpy/kJ mol^{-1}	Ions	Effect of ionic radius of halide ion
NaCl	−787	(+)(−)	As the ionic radius increases: • the charge density decreases • the attraction between ions decreases • the lattice energy becomes less negative
NaBr	−751	(+)(−)	
NaI	−705	(+)(−)	

Effect of ionic charge

The strongest ionic lattices contain **small**, **highly charged ions**.

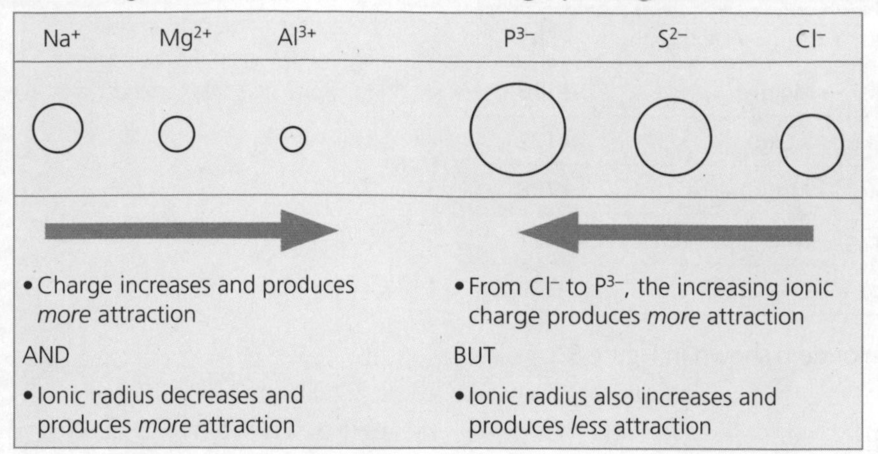

Examiner's tip

Note that lattice energy has a negative value. When describing lattice enthalpies you should use the term 'becomes less/more negative', *not* 'becomes bigger/smaller'.

Enthalpy change of hydration

Revised

The concept of a Born–Haber cycle can be extended to provide a partial explanation of the solubility of substances in water. To do this, the **enthalpy of hydration** of an ion has to be used.

The **enthalpy of hydration** of an ion is the enthalpy change that occurs when 1 mol of gaseous ions is completely hydrated by water.

It is, therefore, the enthalpy change for the process:

$$X^{n+}(g) \rightarrow X^{n+}(aq)$$

The standard enthalpy of hydration applies to the usual conditions of 25°C and 101 kPa.

In the case of hydration of NaCl the attraction is either between the cation, Na^+, and the oxygen atom of a water molecule or between the anion, Cl^-, and the hydrogen atom of the water molecule. This occurs because of the dipoles present in water, which you should be familiar with from the AS course.

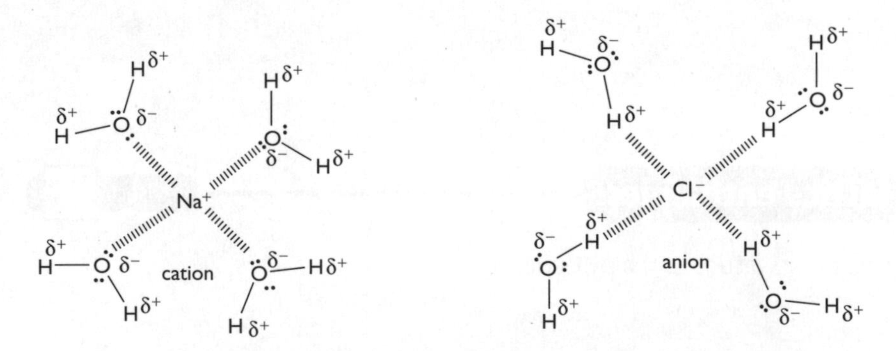

Figure 5.3 Hydration of sodium chloride

Like lattice enthalpy, the enthalpy change of hydration depends on the ionic radius and the size of the charge of the ion. As with lattice enthalpy, the greater the charge density, the greater the attraction.

The enthalpy of solution of a compound is the enthalpy change when 1 mol of that compound dissolves completely in excess water.

Values of lattice enthalpies and enthalpies of hydration relate to the **enthalpy of solution**. A typical enthalpy cycle for sodium chloride is shown opposite.

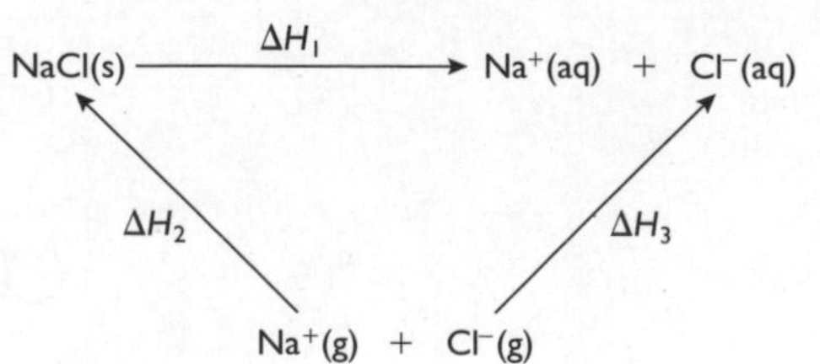

Applying Hess's law:

- ΔH_1 is the enthalpy of solution.
- ΔH_2 is the lattice enthalpy of sodium chloride ($-781\,kJ\,mol^{-1}$).
- ΔH_3 is the enthalpy of hydration of the sodium ion ($-418\,kJ\,mol^{-1}$) + the enthalpy of hydration of the chloride ion ($-338\,kJ\,mol^{-1}$) = $-756\,kJ\,mol^{-1}$

$$\Delta H_2 + \Delta H_1 = \Delta H_3$$

$$\Delta H_1 = \Delta H_3 - \Delta H_2 = -756 + 781 = +25\,kJ\,mol^{-1}$$

The dissolving of sodium chloride is endothermic, yet sodium chloride dissolves readily in water at 25°C. This suggests that there is some other factor that is encouraging the dissolving to take place. This is an energy-related quantity called **entropy,** which is discussed in the next section.

Examiner's tip

This topic could be examined by providing you with a Born–Haber diagram and asking you to deduce the value of any one step. For sodium chloride, it would look like the diagram shown below:

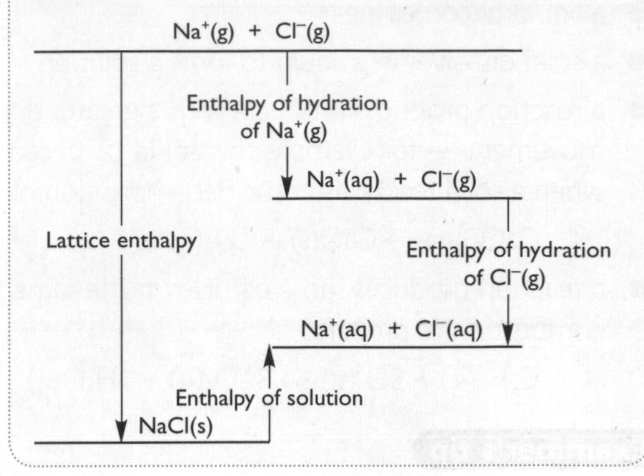

Enthalpy and entropy

Entropy

Revised

When a reaction occurs energy is either absorbed (endothermic) or released (exothermic) in the form of heat. In addition, some energy is either absorbed or released as a result of the re-distribution of the particles when the products are formed. The quantity of energy depends largely on the physical state of the substances and on the temperature.

Entropy is the term used to measure this quantity of energy. It is given the symbol, S.

Solids are more ordered than liquids, and liquids are more ordered than gases. It follows that most energy (entropy) is required to hold a solid in its ordered state. The particles of a gas are less constrained than in a liquid and energy (entropy) is not used because their freedom of movement is not restricted.

The enthalpy change for the melting of ice to water at 0°C is:

$$H_2O(s) \rightarrow H_2O(l) \qquad \Delta H = +6.02\,kJ\,mol^{-1}$$

The conversion therefore looks not to be possible — yet ice does melt at 0°C. This is because, as melting occurs, the change in entropy releases sufficient energy to counteract the positive enthalpy. The energy required to hold the rigid structure of the ice in place is released as the less constrained molecules of water are produced.

Two important differences between enthalpy and entropy are shown in Table 5.2.

Table 5.2 Differences between enthalpy and entropy

Enthalpy	Entropy
If enthalpy is released when a reaction occurs, ΔH is negative	If entropy increases when a reaction occurs, ΔS is positive
The energy unit is usually kilojoules, kJ	The energy unit is usually joules, J

Entropy always increases (ΔS is positive) when there is a greater opportunity for energy to be spread out as a result of a change. The system becomes more disordered. It follows that entropy increases when:

- a solid becomes a liquid
- a liquid becomes a gas
- a solid dissolves in a liquid to form a solution
- a reaction produces products with a greater degree of freedom of movement — for example, this could be because a gas is produced when a solid reacts, as in the decomposition of calcium carbonate:

$$CaCO_3(s) \rightarrow CaO(s) + CO_2(g)$$

- a reaction produces more particles in the same state, as in the combustion of propane:

$$C_3H_8(g) + 5O_2(g) \rightarrow 3CO_2(g) + 4H_2O(g)$$

Examiner's tip

You will not be asked to explain what 'entropy' is because this is a very difficult concept about which even experts disagree. You may be asked to predict whether entropy increases in a reaction — just use the state symbols as a guide. Compare the number of moles of gas, liquid or solid on each side of the equation.

Now test yourself

1 For each of the following reactions, predict whether the entropy change will be positive or negative:

(a) $H_2O(g) \rightarrow H_2O(s)$

(b) $NaOH(s) \rightarrow NaOH(aq)$

(c) $2Mg(s) + O_2(g) \rightarrow 2MgO(s)$

(d) $2SO_2(g) + O_2(g) \rightarrow 2SO_3(g)$

Answers on p. 105

Calculating entropy changes

Revised

Calculations to determine ΔS are similar to calculations for ΔH, although it must be remembered that if ΔS is positive, it means that entropy is increased.

The change in entropy, ΔS, can be calculated using the formula:

$\Delta S = \Sigma$(entropy of products) – Σ(entropy of reactants)

Example

Calculate the entropy change for the following reaction under standard conditions:

$$3O_2(g) \rightarrow 2O_3(g)$$

	$S^{\ominus}$/J mol^{-1} K^{-1}
O_3	238.8
O_2	205

Answer

$\Delta S = \Sigma$(entropy of products) – Σ(entropy of reactants)

$\Delta S = 2 \times (238.8) - 3 \times (205)$

$= -137.4\,J\,mol^{-1}\,K^{-1}$

You should be able to anticipate the sign of ΔS by the fact that 3 mol of O_2 gas have been converted to 2 mol of O_3 gas. When a reaction produces fewer particles in the same state, ΔS is negative.

Now test yourself

2 Substances A, B and C are iodine, ammonia and methanol, but not necessarily in that order. Given the following entropies, identify which substance corresponds to which letter. Explain your answer.

A: $192.5\,J\,mol^{-1}\,K^{-1}$

B: $58.4\,J\,mol^{-1}\,K^{-1}$

C: $127.2\,J\,mol^{-1}\,K^{-1}$

3 Calculate the entropy change when sodium reacts with oxygen.

$S^{\ominus}$(sodium) = $51.0\,J\,mol^{-1}\,K^{-1}$

$S^{\ominus}$(oxygen) = $102.5\,J\,mol^{-1}\,K^{-1}$

$S^{\ominus}$(sodium oxide) = $72.8\,J\,mol^{-1}\,K^{-1}$

Answers on p. 105

Free energy

Revised

The change in the entropy of a reaction can be combined with the change in enthalpy to provide an answer to the question of whether a chemical reaction is feasible.

A new term must be introduced. This is **free energy** (or Gibbs free energy), which is given the symbol G.

The free energy change of a reaction relates to the enthalpy and entropy changes by the equation:

$$\Delta G = \Delta H - T\Delta S$$

ΔG provides a definite answer as to whether a given reaction is feasible. If ΔG is negative, the reaction is feasible; if ΔG is positive, the reaction is not feasible.

Examiner's tip

ΔG is temperature dependent *only* if ΔH and ΔS have the same signs. If ΔH and ΔS are both positive, the reaction is feasible at high temperatures but if ΔH and ΔS are both negative, the reaction will be feasible only at low temperatures.

	ΔH	ΔS	Feasibility of reaction
Signs of ΔH and ΔS are different	+	–	Never
	–	+	Always
Signs of ΔH and ΔS are the same	+	+	Only at high temperatures
	–	–	Only at low temperatures

Typical mistake

When using the expression $\Delta G = \Delta H - T\Delta S$, you must remember that whereas the units of ΔG and ΔH are kJ mol^{-1}, ΔS is measured in J mol^{-1} K^{-1}. Therefore ΔS has to be converted into kJ mol^{-1} K^{-1} by dividing by 1000.

Equilibrium

Revised

When $\Delta G = 0$, the system is at equilibrium and $\Delta H = T\Delta S$.

The enthalpy change for the melting of ice at 0° C is:

$$H_2O(s) \rightarrow H_2O(l) \quad \Delta H = +6.02\,\text{kJ mol}^{-1}$$

At 0°C (273 K), $\Delta G = 0$, so

$$\Delta S = \frac{\Delta H}{T}$$

Hence, the entropy change when ice melts is:

$$\Delta S = \frac{\Delta H}{T} = \frac{6.02\text{ kJ mol}^{-1}}{273\text{ K}}$$

Entropy is measured in J mol^{-1} K^{-1}, so

$$\Delta S = \frac{6020\text{ J mol}^{-1}}{273\text{ K}} = 22.0\,\text{J mol}^{-1}\,\text{K}^{-1}$$

For a chemical reaction, the values of ΔH and ΔS must first be calculated. Then, the value of the temperature, T, for which ΔG is zero can be established. It should be noted that equilibrium can never be achieved for a reaction in which ΔH is negative and ΔS is positive (because the reaction is always feasible) or in which ΔH is positive and ΔS is negative (because the reaction is never feasible). When ΔH and ΔS have the same sign, it is possible to find the equilibrium temperature, noting that:

as $\Delta G = 0$, $\Delta H = T\Delta S$, so $T = \dfrac{\Delta H}{\Delta S}$

Calculating free energy changes

If tables of information are provided, then calculating ΔG is similar to the process of calculating ΔH.

Example

Use the data in the table to calculate the temperature at which the reaction $2NO(g) + O_2(g) \rightarrow 2NO_2(g)$ reaches equilibrium.

	$\Delta H_f^\ominus$/kJ mol^{-1}	$S^\ominus$/J mol^{-1} K^{-1}
NO(g)	90.4	210.5
O_2(g)	0	204.9
NO_2(g)	33.2	240.0

Answer

$\Delta H = \Sigma$(enthalpy of products) − Σ(enthalpy of reactants)

$\Delta S = \Sigma$(entropy of products) − Σ(entropy of reactants)

At equilibrium, $\Delta G = 0$ so $T = \frac{\Delta H}{\Delta S}$

ΔH for the reaction is $(2 \times 33.2) - (2 \times 90.4) = -114.4\,\text{kJ mol}^{-1}$

$\Delta S = (2 \times 240.0) - ((2 \times 210.5) + 204.9) = -145.9\,\text{J mol}^{-1}\,\text{K}^{-1}$

Therefore, (remembering to convert the energy unit of ΔS from J into kJ):

$$T = \frac{-114.4}{-0.1459} = 784\,\text{K or } 511°\text{C}$$

Electrode potentials and fuel cells

Ionic equations

Revised

To write ionic equations correctly, it is essential to balance *both* symbol and charge. State symbols should always be included.

Typical mistake

Students are aware that symbols have to be balanced, but they often ignore charge, which must also be balanced. For example, when asked to balance the equation:

$...IO_3^-(aq) + ...I^-(aq) + ...H^+(aq) \rightarrow ...I_2(s) + ...H_2O(l)$, the most common response is:

$...IO_3^-(aq) + ...I^-(aq) + 6H^+(aq) \rightarrow ...I_2(s) + 3H_2O(l)$

which balances the symbols but not the charges. The correct response is:

$IO_3^-(aq) + 5I^-(aq) + 6H^+(aq) \rightarrow 3I_2(s) + 3H_2O(l)$ which balances both.

Examiner's tip

When you have balanced an equation *always* double-check to make sure that the charges balance as well as the symbols. In the equation illustrated in 'Typical mistakes', each side of the equation has a net charge of zero, hence it is balanced.

The reaction between an acid and a base to produce a salt and water can be represented by an ionic equation. For example, when an aqueous hydroxide reacts with an acid, the ionic equation is:

$$H^+(aq) + OH^-(aq) \rightarrow H_2O(l)$$

Both sides of the equation have a net charge of zero.

When the base magnesium oxide reacts with an acid the ionic equation is:

$$MgO(s) + 2H^+(aq) \rightarrow Mg^{2+}(aq) + H_2O(l)$$

In an ionic solid such as MgO, the ions are not free to move, so they are not written as separate ions.

Both sides of the equation have a net charge of 2+.

The reaction between an acid and a carbonate to produce a salt, carbon dioxide and water can be represented by an ionic equation:

$$CO_3^{2-}(aq) + 2H^+(aq) \rightarrow CO_2(g) + H_2O(l)$$

Both sides of the equation have a net charge of zero.

Now test yourself

4 Write ionic equations for each of the following reactions:
- (a) Aqueous potassium carbonate and nitric acid
- (b) Solid calcium carbonate and hydrochloric acid
- (c) Precipitation of calcium carbonate from aqueous calcium chloride and aqueous sodium carbonate
- (d) Neutralisation of aqueous calcium hydroxide and hydrochloric acid
- (e) Precipitation of copper (II) hydroxide from aqueous copper nitrate and aqueous potassium hydroxide
- (f) Zinc oxide and nitric acid

Answers on p. 105

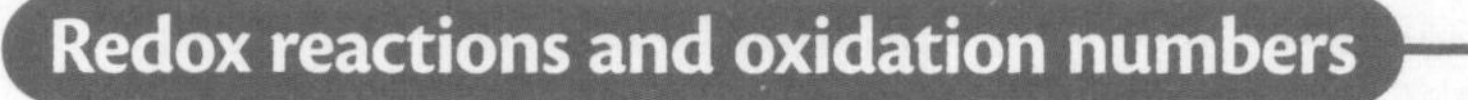

Redox reactions and oxidation numbers

Revised

The displacement reaction between chlorine and bromide is an example of a redox reaction that you met in Unit F321 at AS.

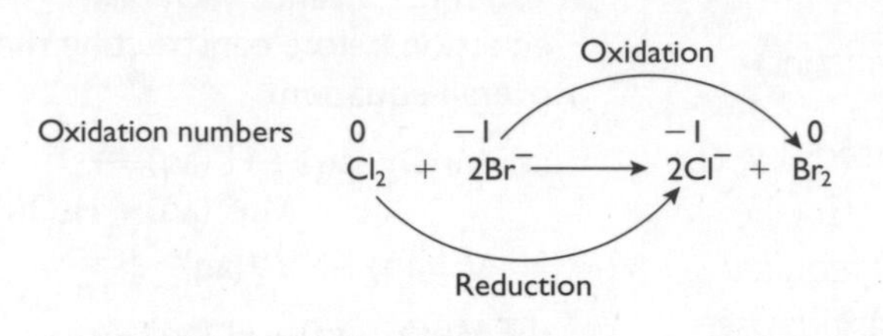

Remember that **o**xidation **i**s **l**oss, **r**eduction **i**s **g**ain ('OILRIG').

Oxidation number is a convenient way of identifying quickly whether a substance has undergone either oxidation or reduction. In order to work out the oxidation number, you must first learn a few simple rules.

Table 5.3 Rules for working out oxidation number

	Rule	Example
1	All elements in their natural state have an oxidation number of zero	H_2, oxidation number of H is zero
2	The oxidation numbers in any molecule always add up to zero	H_2O, sum of oxidation numbers is zero
3	The oxidation numbers of any ion always add up to the charge on the ion	SO_4^{2-}, sum of oxidation numbers is −2
4	Elements in groups 1, 2 and 3 have oxidation numbers of +1, +2 and +3 respectively	NaCl: Na is +1 $MgCl_2$: Mg is +2 $AlCl_3$: Al is +3
5	Fluorine is always −1	HF: F is −1
6	Hydrogen is usually +1	H_2O: H is +1
7	Oxygen is usually −2	H_2O: O is −2
8	Transition elements have no fixed oxidation number	Variable oxidation numbers (e.g. Fe: +2, +3)

If you make sure you apply these rules rigidly in the sequence indicated it should be relatively simple to deduce any oxidation number.

Consider the reaction of zinc and aqueous copper sulfate:

$$Zn(s) + CuSO_4(aq) \rightarrow ZnSO_4(aq) + Cu(s)$$

It may be helpful to write the oxidation numbers above each element in the equation:

oxidation number: → 0 +2 +6 −2 +2+6 −2 0

$$Zn(s) + CuSO_4(aq) \rightarrow ZnSO_4(aq) + Cu(s)$$

In any redox reaction the oxidation number of one element increases and the oxidation number of a second element decreases. The oxidation number of zinc increases from 0 to +2. The oxidation number of copper decreases from +2 to 0. The ionic equation is:

$$Zn(s) + Cu^{2+}(aq) \rightarrow Zn^{2+}(aq) + Cu(s)$$

The zinc is oxidised. Each zinc atom loses two electrons and becomes a Zn^{2+} ion. The electrons are taken up by a Cu^{2+} ion, which is reduced to a copper atom, Cu.

The ionic half-equations are:

$Zn(s) \rightarrow Zn^{2+}(aq) + 2e^-$ oxidation (loss of electrons)

$Cu^{2+}(aq) + 2e^- \rightarrow Cu(s)$ reduction (gain of electrons)

It is also possible to use ionic half-equations to construct a full ionic equation. When Cu(s) is added to aqueous $AgNO_3$(aq), Cu^{2+}(aq) and Ag(s) are formed. The ionic half-equations are:

$Cu(s) \rightarrow Cu^{2+}(aq) + 2e^-$ oxidation (loss of electrons)

$Ag^+(aq) + e^- \rightarrow Ag(s)$ reduction (gain of electrons)

In any pair of ionic half-equations, the number of electrons released (by oxidation) is the same as the number required for the reduction. Cu(s) supplies two electrons as it is oxidised to Cu^{2+}(aq). Each Ag^+(aq) requires only one electron to be reduced to Ag(s). Therefore, it is necessary to use two Ag^+(aq) for each Cu(s):

$Cu(s) \rightarrow Cu^{2+}(aq) + \mathbf{2e^-}$

$2Ag^+(aq) + \mathbf{2e^-} \rightarrow 2Ag(s)$

The overall equation is obtained by adding the two half-equations together, excluding the electrons:

$2Ag^+(aq) + Cu(s) \rightarrow 2Ag(s) + Cu^{2+}(aq)$

The overall equation can now be written:

$2AgNO_3(aq) + Cu(s) \rightarrow 2Ag(s) + Cu(NO_3)_2(aq)$

Now test yourself

5 Use each of the following pairs of half-equations to construct an overall equation for the reaction. You must balance each half-equation before constructing the overall equation.

(a) $MnO_4^-(aq) + H^+(aq) \rightarrow Mn^{2+}(aq) + H_2O(l)$

$V^{2+}(aq) \rightarrow V^{3+}(aq)$

(b) $MnO_4^-(aq) + H^+(aq) \rightarrow Mn^{2+}(aq) + H_2O(l)$

$V^{2+}(aq) + H_2O \rightarrow VO_3^-(aq) + H^+(aq)$

(c) $Cr_2O_7^{2-}(aq) + H^+(aq) \rightarrow Cr^{3+}(aq) + H_2O(l)$

$SO_2(aq) + H_2O(l) \rightarrow SO_4^{2-}(aq) + H^+(aq)$

(d) $NO_3^-(aq) + H^+(aq) \rightarrow NO(g) + H_2O(l)$

$Cu(s) \rightarrow Cu^{2+}(aq)$

Answers on p. 105, 106

Electrode potentials

Revised

The **standard electrode potential** is the potential difference (the difference in voltage) between one half-cell (e.g. a metal in contact with its metal ions) and the standard hydrogen electrode, when measured under standard conditions.

The **standard cell potential** is the voltage formed when two half-cells are connected. It is measured using a voltmeter of high resistance under standard conditions.

Standard conditions are: temperature = 298 K, pressure =101 kPa, concentration = 1.0 mol dm^{-3}.

A diagram representing the standard hydrogen electrode is shown in Figure 5.4.

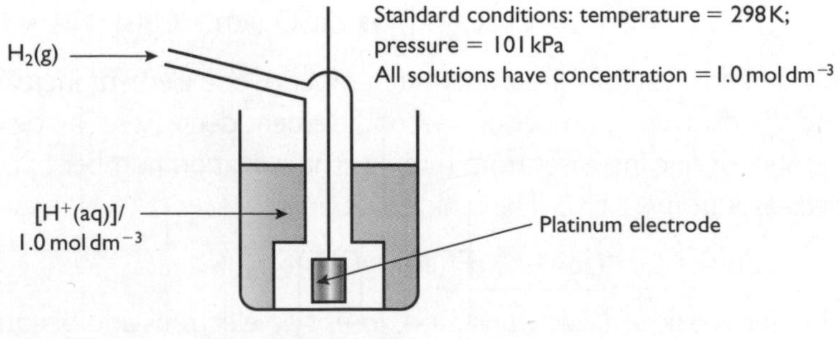

Standard conditions: temperature = 298 K; pressure = 101 kPa
All solutions have concentration = 1.0 mol dm^{-3}

$H_2(g)$

$[H^+(aq)]$/ 1.0 mol dm^{-3}

Platinum electrode

Figure 5.4 A standard hydrogen electrode

To provide a better surface for the hydrogen, the platinum electrode is usually coated with very finely divided platinum known as 'platinum black'. This cell is connected via an external circuit and through a salt bridge to the other cell. The voltage measured gives the electrode potential of this cell compared with the half-reaction:

$2H^+(aq) + 2e^- \rightarrow H_2(g)$

which is given the arbitrary value of zero.

Measuring standard electrode potentials

(1) Metals

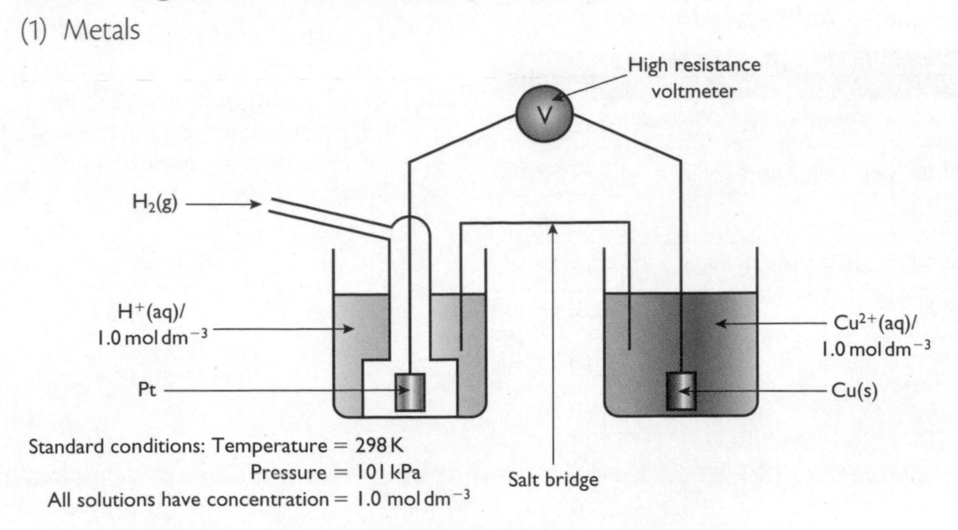

(2) Non-metals/ions of the same element in different oxidation states

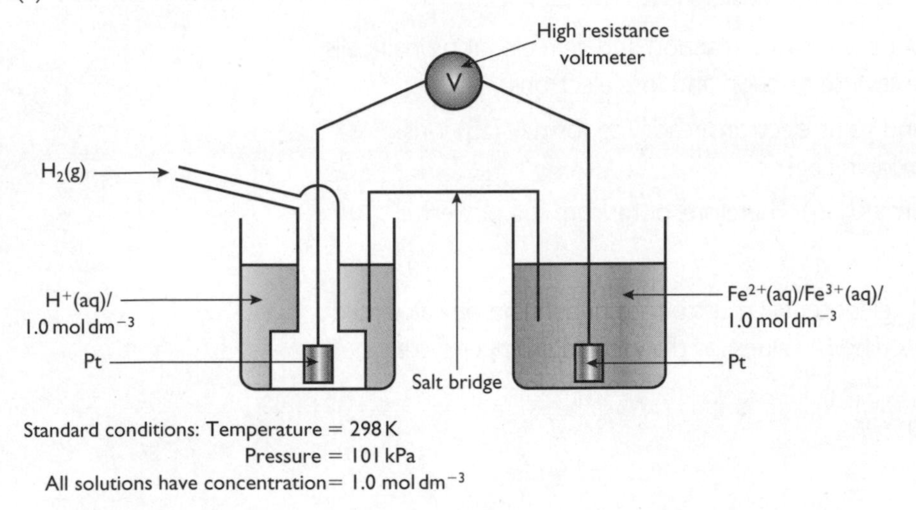

The salt bridge is made of a porous material soaked in a saturated solution of KNO_3. The salt bridge completes the circuit without mixing the solutions by allowing the passage of ions.

Typical mistake

If asked to draw a sketch to show how to measure the electrode potential for Fe^{2+}/Fe^{3+} many candidates incorrectly draw:

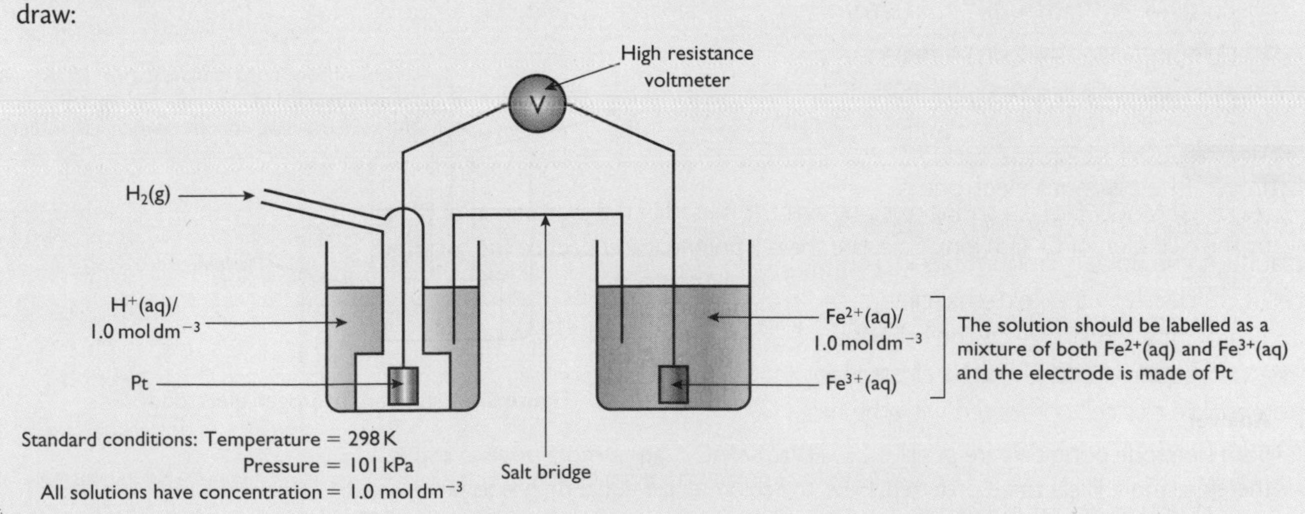

Some common cell potentials are listed in the table below.

Table 5.4 Common cell potentials

	Half-cell	**$E^{\ominus}$/volts**
A	$F_2(g) + 2e^- \rightleftharpoons 2F^-(aq)$	+2.87
B	$MnO_4^-(aq) + 8H^+(aq) + 5e^- \rightleftharpoons Mn^{2+}(aq) + 4H_2O(l)$	+1.52
C	$Cl_2(g) + 2e^- \rightleftharpoons 2Cl^-(aq)$	+1.36
D	$Cr_2O_7^{2-}(aq) + 14H^+(aq) + 6e^- \rightleftharpoons 2Cr^{3+}(aq) + 7H_2O(l)$	+1.33
E	$Ag^+(aq) + e^- \rightleftharpoons Ag(s)$	+0.80
F	$Fe^{3+}(aq) + e^- \rightleftharpoons Fe^{2+}(aq)$	+0.77
G	$Cu^{2+}(aq) + 2e^- \rightleftharpoons Cu(s)$	+0.34
H	**$2H^+(aq) + 2e^- \rightleftharpoons H_2(g)$**	0.00
I	$Zn^{2+}(aq) + 2e^- \rightleftharpoons Zn(s)$	−0.76
J	$K^+(aq) + e^- \rightleftharpoons K(s)$	−2.92

Half-cells with positive $E^{\ominus}$ (A–G) favour the forward reaction and gain electrons; half-cells with a negative $E^{\ominus}$ (I and J) favour the reverse reaction and lose electrons.

- $F_2(g)$ has the highest positive $E^{\ominus}$ and gains electrons readily to form $F^-(aq)$ ions. Therefore, fluorine is a powerful oxidising agent.
- K(s) loses an electron readily to form $K^+(aq)$. Therefore, potassium is a powerful reducing agent.

For a reaction to proceed, the overall cell potential must be positive. It can be calculated by using the appropriate two half-cells. The $E^{\ominus}$ values for the zinc and copper systems are:

$$Zn^{2+}(aq) + 2e^- \rightleftharpoons Zn(s) \quad -0.76\,V$$
$$Cu^{2+}(aq) + 2e^- \rightleftharpoons Cu(s) \quad +0.34\,V$$

which indicates that the $Cu^{2+}(aq)$ favours the forward reaction (and moves to the right) while the Zn(s) favours the reverse reaction (and moves to the left). The overall cell potential can be calculated using the equation:

$$E_{cell} = E_{right} - E_{left} = (+0.34) - (-0.76) = 1.10\,V$$

Each half cell can now be rewritten as:

$$Zn(s) \rightarrow Zn^{2+}(aq) + 2e^- \quad +0.76\,V$$
$$Cu^{2+}(aq) + 2e^- \rightarrow Cu(s) \quad +0.34\,V$$
$$\text{Cell potential} = +1.10\,V$$

Example

Acidified $MnO_4^-(aq)$ is a strong oxidising agent. It is used in the preparation of $Cl_2(g)$ by the oxidation of $Cl^-(aq)$ ions. Calculate the cell potential and deduce the balanced equation.

$$MnO_4^-(aq) + 8H^+(aq) + 5e^- \rightleftharpoons Mn^{2+} + 4H_2O(l) \qquad E^{\ominus} = +1.52\,V$$
$$Cl_2(g) + 2e^- \rightleftharpoons 2Cl^-(aq) \quad E^{\ominus} = +1.36\ V$$

Answer

Both electrode potentials are positive, but $H^+(aq)/MnO_4^-(aq)$ is more positive and is, therefore, more likely to be preferred. Here, the chlorine half-equation has to be reversed:

$$MnO_4^-(aq) + 8H^+(aq) + 5e^- \rightleftharpoons Mn^{2+} + 4H_2O(l) \qquad E^{\ominus} = +1.52\,V$$
$$2Cl^-(aq) \rightarrow Cl_2(g) + 2e^- \quad E^{\ominus} = -1.36\,V$$

Using $E_{cell} = E_{right} - E_{left} = (+1.52) - (+1.36)$

Cell potential = +0.16 V

Both half-equations must have the same number of electrons. The MnO_4^- half-equation is multiplied by 2 to give $10e^-$; the Cl^- half-equation is multiplied by 5.

$2MnO_4^-(aq) + 16H^+(aq) + 10e^- \rightarrow 2Mn^{2+}(aq) + 8H_2O(l)$

$10Cl^-(aq) \rightarrow 5Cl_2(g) + 10e^-$

The overall equation is:

$2MnO_4^-(aq) + 16H^+(aq) + 10Cl^-(aq) \rightarrow 2Mn^{2+}(aq) + 8H_2O(l) + 5Cl_2(g)$

Examiner's tip

When balancing an equation always check that it is balanced for both symbol and charge. The reaction between MnO_4^- and Cl^- in the example has a net charge of +4 on each side of the equation. Hence, it is balanced.

Now test yourself

6 $Mg^{2+}(aq) + 2e^- \rightleftharpoons Mg(s)$ $E^\ominus = -2.37$ V

$Zn^{2+}(aq) + 2e^- \rightleftharpoons Zn(s)$ $E^\ominus = -0.76$ V

$Sn^{4+}(aq) + e^- \rightleftharpoons Sn^{2+}(aq)$ $E^\ominus = +0.15$ V

$I_2(aq) + 2e^- \rightleftharpoons 2I^-(aq)$ $E^\ominus = +0.54$ V

$Fe^{3+}(aq) + e^- \rightleftharpoons Fe^{2+}(aq)$ $E^\ominus = +0.77$ V

$Br_2(aq) + 2e^- \rightleftharpoons 2Br^-(aq)$ $E^\ominus = +1.09$ V

Use the standard electrode potentials, $E^\ominus$, listed above to calculate the cell potential for each of the following pairs of half-cells:

(a) $Mg(s)/Mg^{2+}(aq)$ and $Zn(s)/Zn^{2+}(aq)$
(b) $Sn^{4+}(aq)/Sn^{2+}(aq)$ and $Fe^{3+}(aq)/Fe^{2+}(aq)$
(c) $I_2(aq)/2I^-(aq)$ and $Br_2(aq)/2Br^-(aq)$
(d) $Zn(s)/Zn^{2+}(aq)$ and $I_2(aq)/2I^-(aq)$
(e) $Sn^{4+}(aq)/Sn^{2+}(aq)$ and $Br_2(aq)/2Br^-(aq)$

Answers on p. 106

The effect of concentration on the feasibility of reactions

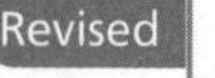

A positive cell potential indicates that a reaction is feasible. However, it gives no indication of how fast a reaction will occur. The cell potential of the reaction between $H^+(aq)/MnO_4^-(aq)$ and Cl^- is only +0.16 V, but the reaction takes place quickly despite the low overall potential.

The cell potentials are calculated assuming standard conditions. According to le Chatelier's principle, if the concentration of one component in a half-cell is changed, the equilibrium will move to minimise the effect of the change.

Consider the equilibrium:

$$Fe^{3+}(aq) + e^- \rightleftharpoons Fe^{2+}(aq) \quad E^\ominus = +0.77\text{ V}$$

If the concentration of $Fe^{2+}(aq)$ is reduced, the equilibrium position will move to the right, causing the value of the electrode potential to increase. If the concentration of $Fe^{3+}(aq)$ is reduced, the equilibrium will move to the left and the value of $E^\ominus$ will decrease. In each case, an extremely large change would be required to make any noticeable difference. As a general rule of thumb, a tenfold change in concentration changes the electrode potential of a half-reaction by 0.06 V or less.

Storage cells

Revised

Storage cells are commonly referred to as batteries. Electrode potential can be used to predict the possible voltage of a battery.

Commercial batteries do not usually contain liquid. The cells usually contain pastes (the electrolyte) that surround the electrodes. An 'alkaline battery' has a cathode made from graphite and manganese(IV) oxide and an anode made of either zinc or nickel-plated steel. The electrolyte is potassium hydroxide.

Examiner's tip

There is no need to remember the details or construction of any particular cell, but you may be asked in an exam to interpret provided data.

The reactions that take place are:

At the anode: $Zn + 2OH^- \rightarrow ZnO + H_2O + 2e^-$

At the cathode: $2MnO_2 + H_2O + 2e^- \rightarrow Mn_2O_3 + 2OH^-$

The overall equation taking place is found by combining the two half-reactions:

$$Zn + 2MnO_2 \rightarrow ZnO + Mn_2O_3$$

Both OH^- and H_2O can be eliminated from the equation because their concentrations should remain constant. (In practice, some loss does occur.)

The overall voltage is about 1.5 V and is a combination of the electrode potentials of the two half-reactions.

Another battery often encountered is the rechargeable nickel–cadmium (Ni–Cd) cell. While it is supplying electricity the reactions taking place are:

At the anode: $Cd + 2OH^- \rightarrow Cd(OH)_2 + 2e^-$

At the cathode: $2NiO(OH) + 2H_2O + 2e^- \rightarrow 2Ni(OH)_2 + 2OH^-$

The electrolyte is potassium hydroxide.

The overall reaction is:

$$Cd + 2NiO(OH) + 2H_2O \rightarrow Cd(OH)_2 + 2Ni(OH)_2$$

The electrolyte, OH^-, cancels out. A voltage of around 1.2 V can be obtained.

The battery can be recharged by applying an external voltage that reverses the reactions shown above. A disadvantage of this type of battery is that cadmium is toxic and care needs to be taken when disposing of the batteries.

Many other storage cells are made and the intense research into the development of batteries for use in vehicles has led to a number of different constructions. A disadvantage of all storage cells is that the reactants are used up and the overall voltage is not constant.

Fuel cells

Revised

A fuel cell produces electrical power from the chemical reaction of a fuel (for example hydrogen, hydrocarbons or alcohols) with oxygen. The fuel cell operates like a conventional storage cell, except that the fuels are supplied externally as gases. The cell will therefore operate more or less indefinitely so long as the fuel supply is maintained and the cell voltage remains constant.

The hydrogen/oxygen fuel cell is used widely and illustrates the principles behind fuel cells in general.

The electrodes are made of a material such as a titanium sponge coated in platinum. The electrolyte is an acid or alkaline membrane that allows ions to move from one compartment of the cell to the other. (In other words it acts like a salt bridge.)

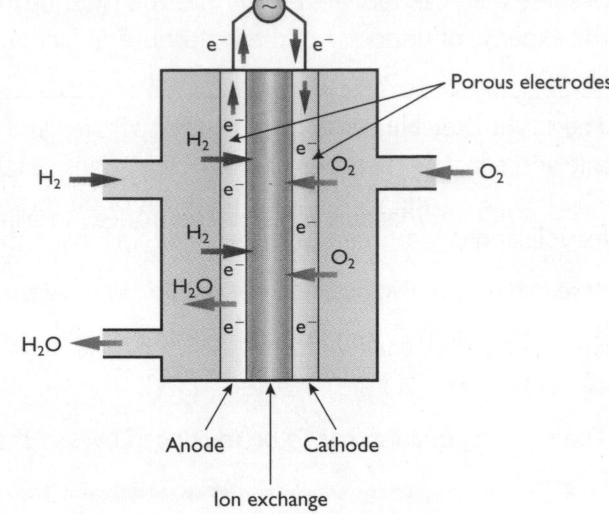

Figure 5.5 A hydrogen/oxygen fuel cell

In an acidic solution, hydrogen is converted to hydrogen ions at the cathode while, at the anode, oxygen reacts with hydrogen ions to make water.

$$H_2(g) \rightleftharpoons 2H^+(aq) + 2e^- \qquad E^\ominus = 0.00\,V$$

$$\tfrac{1}{2}O_2(g) + 2H^+(aq) + 2e^- \rightleftharpoons H_2O(l) \qquad E^\ominus +1.23\,V$$

The overall reaction is:

$$H_2(g) + \tfrac{1}{2}O_2(g) \rightarrow H_2O(l)$$

The voltage produced is 1.23 V.

In alkaline solution, hydrogen reacts with hydroxide ions to form water and oxygen reacts with water to form hydroxide ions:

$$H_2(g) + 2OH^-(aq) \rightleftharpoons 2H_2O(l) + 2e^- \qquad E^\ominus = +0.83\,V$$

$$\tfrac{1}{2}O_2(g) + H_2O(l) + 2e^- \rightleftharpoons 2OH^-(aq) \qquad E^\ominus = +0.40\,V$$

The overall reaction is:

$$H_2(g) + \tfrac{1}{2}O_2(g) \rightarrow H_2O(l)$$

and the voltage produced is again 1.23 V.

Hydrogen–oxygen fuel cells offer an alternative to the use of fossil fuels (petrol or diesel) that will eventually run out. They do not produce products that are pollutants, for example carbon monoxide, carbon dioxide and oxides of nitrogen. They are relatively light and are more efficient than engines that use fossil fuels.

However, there are drawbacks and problems that have to be overcome. Hydrogen is a gas and it is dangerously explosive. Possible solutions include:

- compressing the gas until it liquefies
- adsorbing it onto the surface of a suitable solid material
- absorbing it into a suitable material

A number of transition metal alloys have been tried — for example, an alloy of iron and titanium, which absorbs hydrogen quite well but is too heavy to be practical. Another possibility is to use carbon nanotubes. These are tiny lightweight structures containing arrangements of carbon atoms that have small, cylindrical pores which capture the hydrogen. This material is light enough but needs further investigation before it could be considered to be a commercial proposition.

Many metals can also absorb hydrogen to form metal hydrides, which can then, under the right conditions, release hydrogen. Particular interest has been shown in the use of lighter metals — for example magnesium, which forms the hydride MgH_2. A problem with this is that the hydride needs to be quite hot before the hydrogen is released. In addition, the manufacture of these materials requires energy and this must be included in assessing their viability, as must the expense of disposing of them after use. In fact, hydrides tend to have a rather limited lifetime.

The major issue concerning the use of hydrogen as a fuel is safety. Research is continuing into safe and effective methods of production, transportation and storage of hazardous materials. These issues, along with finding methods of production of hydrogen that are economically feasible and environmentally acceptable, will shape the possible future use of hydrogen as a fuel.

Research is also being carried out on alcohol–oxygen fuel cells. The reactions involved are:

$$CH_3OH(l) + H_2O(l) \rightarrow 6H^+(aq) + CO_2(g) + 6e^-$$
$$\tfrac{1}{2}O_2(g) + 2H^+(aq) + 2e^- \rightarrow H_2O(l)$$

The second equation has to be multiplied by 3 so that the electrons cancel, such that the net reaction is:

$$CH_3OH(l) + \tfrac{3}{2}O_2(g) \rightarrow 2H_2O(l) + CO_2(g)$$

The advantage of using methanol in place of hydrogen is that methanol is a liquid and is easier and safer to store and transport. The disadvantage of using methanol in place of hydrogen is that $CO_2(g)$ is produced as a waste product.

Check your understanding

1 Use the data provided to:

(a) construct a Born–Haber cycle for potassium oxide

(b) calculate the lattice enthalpy of potassium oxide

Enthalpy of atomisation of potassium = $+89.5\,kJ\,mol^{-1}$
First ionisation energy of potassium = $+420.0\,kJ\,mol^{-1}$
Enthalpy of atomisation of oxygen = $+249.4\,kJ\,mol^{-1}$
First electron affinity of oxygen = $-141.4\,kJ\,mol^{-1}$
Second electron affinity of oxygen = $+790.8\,kJ\,mol^{-1}$
Enthalpy of formation of potassium oxide = $-361.5\,kJ\,mol^{-1}$

2 Use the enthalpies provided to calculate:

(a) the enthalpy of solution of silver chloride

(b) the enthalpy of hydration of iodide ions, $I^-(aq)$

Enthalpy of hydration of Ag^+ = $-464.4\,kJ\,mol^{-1}$
Enthalpy of hydration of Cl^- = $-384.1\,kJ\,mol^{-1}$
Enthalpy of solution of AgI = $+96.9\,kJ\,mol^{-1}$
Lattice enthalpy of AgCl = $-890\,kJ\,mol^{-1}$
Lattice enthalpy of AgI = $-867\,kJ\,mol^{-1}$

(c) Compare your answer to part (b) with the enthalpy of hydration of $Cl^-(aq)$ ($-384.1\,kJ\,mol^{-1}$). Explain your answer.

(d) Compare the relative solubilities of AgCl and AgI.

3 Dental amalgam contains about 40% mercury (Hg) combined with an alloy that is made largely of silver and tin. (Small amounts of copper and zinc are also present.) A number of redox reactions are possible with the amalgam as an electrode and saliva in the mouth as the electrolyte.

Two examples are:

$Hg^+ + e^- \rightleftharpoons$ Ag/Hg (amalgam) $\quad E = +0.85\,V$

$Sn^{2+} + 2e^- \rightleftharpoons$ Sn/Hg (amalgam) $\quad E = -0.13\,V$

If a piece of aluminium foil is bitten by teeth containing an amalgam, an unpleasant sharp pain is experienced. This results from a temporary cell being set up between the amalgam and the aluminium:

$Al^{3+}(aq) + 3e^- \rightleftharpoons Al(s) \quad E = -1.66\,V$

Describe what happens in the cell and why it results in pain being felt.

Answers on pp. 106, 107

Exam practice

1 (a) (i) Explain what is meant by the term lattice enthalpy. [2]

(ii) Write an equation to show what is meant by the lattice enthalpy of magnesium chloride. [2]

(b) Use the Born–Haber cycle below to answer the questions that follow.

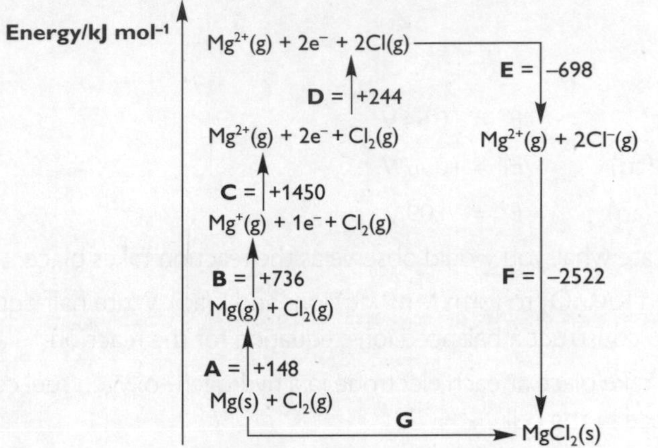

(i) Identify which step represents the second ionisation energy of magnesium. [1]

(ii) Write an equation that illustrates the second ionisation energy of magnesium. [1]

(iii) Explain why the enthalpy value for the second ionisation energy of magnesium is about twice the value of the first ionisation energy. [2]

(iv) Write an equation that illustrates the first electron affinity of chlorine. [2]

(v) State the energy, in kJ mol^{-1}, for the first electron affinity of chlorine. [1]

(vi) Calculate the enthalpy of formation of magnesium chloride. [2]

(c) The lattice enthalpy for magnesium bromide is −2440 kJ mol^{-1}. Explain the difference in the values of the lattice enthalpies of magnesium bromide and magnesium chloride. [1]

(d) Magnesium bromide is soluble in water. The enthalpy of hydration of magnesium ions is −1921 kJ mol^{-1} and the enthalpy of hydration of the bromide ion is −336 kJ mol^{-1}. Calculate the enthalpy of solution of magnesium bromide. [2]

(e) Using enthalpies of solution is not always a reliable way of predicting whether a substance will be soluble in water.

(i) What other energy change should be considered in making a prediction of solubility? [1]

(ii) Explain whether this other energy change is likely to suggest that a substance will be more or less soluble in water. [2]

(f) Describe how you could distinguish between aqueous solutions of magnesium bromide and magnesium chloride. State the observations you would make. [3]

2 (a) Use the data below to calculate the standard enthalpy change for the reaction:

$C(s) + CO_2(g) \rightarrow 2CO(g)$

$\Delta H_f^\ominus(CO_2) = -393.7\ kJ\ mol^{-1}$

$\Delta H_f^\ominus(CO) = -110.5\ kJ\ mol^{-1}$ [2]

(b) Use the data below to calculate the standard entropy change for the reaction:

$C(s) + CO_2(g) \rightarrow 2CO(g)$

$S^\ominus(CO_2) = 213.8\ J\ mol^{-1}\ K^{-1}$

$S^\ominus(C) = 5.7\ J\ mol^{-1}\ K^{-1}$

$S^\ominus(CO) = 197.9\ J\ mol^{-1}\ K^{-1}$ [2]

(c) (i) State the relationship between ΔG, ΔH and ΔS. [1]

(ii) Use your answers to (a) and (b) to determine the value of $\Delta G^\ominus$ for the reaction between carbon dioxide and carbon under standard conditions of 298 K and 101 kPa. [2]

(d) Calculate the minimum temperature in °C required for the reaction between carbon dioxide and carbon to become feasible. [3]

3 (a) Draw a diagram to show how to measure the standard electrode potential of the half-cell:

$Fe^{3+}(aq) + e^- \rightleftharpoons Fe^{2+}(aq)$

State the conditions necessary. **[6]**

(b) Use the electrode potentials provided to predict whether, under standard conditions, $Fe^{3+}(aq)$ will be able to react with:

(i) $I^-(aq)$

(ii) $Br^-(aq)$

$I_2(aq) + 2e^- \rightleftharpoons 2I^-(aq)$ $\quad E^\ominus = +0.54\,V$

$Fe^{3+}(aq) + e^- \rightleftharpoons Fe^{2+}(aq)$ $\quad E^\ominus = +0.77\,V$

$Br_2(aq) + 2e^- \rightleftharpoons 2Br^-(aq)$ $\quad E^\ominus = +1.09\,V$

If a reaction is possible, state what you would observe as the reaction takes place. **[5]**

(c) $I^-(aq)$ reacts with acidified $KMnO_4$ to form $Mn^{2+}(aq)$ ions and $I_2(aq)$. Write half-equations for each of these reagents and use them to construct a balanced ionic equation for the reaction. **[3]**

4 (a) Explain the changes that take place at each electrode in a hydrogen–oxygen fuel cell. Give the overall reaction that is taking place in the cell. **[4]**

(b) Under standard conditions, a fuel cell can produce a voltage of 1.23 V. If the cell was used in a vehicle this voltage would be less than 1.23 V. Suggest two reasons why the voltage might be less. **[2]**

(c) State three advantages that using fuel cells in vehicles might have over using petrol as the source of energy. **[3]**

(d) Suggest two ways in which hydrogen might be stored in a vehicle using a fuel cell. **[2]**

Answers and quick quiz 5 online

Online

Examiner's summary

You should now have an understanding of:

- ✔ lattice enthalpy and Born–Haber cycles
- ✔ enthalpies of solution and hydration
- ✔ entropy, and the relationship between free energy, enthalpy and entropy
- ✔ electrode potentials
- ✔ storage cells and fuel cells

6 Transition elements

Transition elements

The **transition elements** occur in period 4 of the periodic table.

A **transition element** is defined as a *d*-block element that forms one (or more) stable ion that has partly filled *d*-orbitals.

The 4*s* sub-shell is at a lower energy level than the 3*d* sub-shell and therefore the 4*s* sub-shell fills before the 3*d* sub-shell. The orbitals in the 3*d* sub-shell are first occupied singly to prevent any repulsion caused by pairing.

The majority of transition elements form ions in more than one oxidation state. When transition elements form ions they do so by losing electrons from the 4*s* orbitals before the 3*d* orbitals. Sc and Zn each form ions in one oxidation state only: Sc^{3+} and Zn^{2+}. The electron configurations of these ions are $[Ar]3d^0$ and $[Ar]3d^{10}$ respectively. Neither fits the definition of a transition element.

Typical mistake

Students usually get the full electron configuration of a transition metal correct. However, when asked for the electron configuration of a transition metal *ion* many incorrectly remove the 3*d* electrons before the 4*s*. The electron configuration of $_{26}Fe^{2+}$ is $1s^22s^22p^63s^23p^63d^6$, *not* $1s^22s^22p^63s^23p^63d^44s^2$.

Table 6.1 Electron configuration of the elements in period 4

Element	Electron configuration
Sc	$[Ar]3d^14s^2$
Ti	$[Ar]3d^24s^2$
V	$[Ar]3d^34s^2$
*Cr	$[Ar]3d^54s^1$
Mn	$[Ar]3d^54s^2$
Fe	$[Ar]3d^64s^2$
Co	$[Ar]3d^74s^2$
Ni	$[Ar]3d^84s^2$
**Cu	$[Ar]3d^{10}4s^1$
Zn	$[Ar]3d^{10}4s^2$

* Chromium has one electron in each orbital of the 4*s* and 3*d* sub-shells giving the configuration as $[Ar]3d^54s^1$, which is more stable than $[Ar]3d^44s^2$.

** Copper has a full 3*d* sub-shell giving the configuration as $[Ar]3d^{10}4s^1$, which is more stable than $[Ar]3d^94s^2$.

Properties of transition elements

Revised

- The transition elements are all metals and therefore they are good conductors of heat and electricity.
- They are denser than other metals. They have smaller atoms than the metals in groups 1 and 2 so the atoms pack together more closely, hence increasing the density.
- They have higher melting and boiling points than other metals. This can be explained by considering the size of their atoms. Within the metallic lattice, ions are smaller than those of the *s*-block metals which results in greater 'free electron density' and hence a stronger metallic bond.
- They have compounds with two or more oxidation states. This is because successive ionisation energies of transition metals increase only gradually. All the transition metals can form an ion of oxidation state +2, representing the loss of the two 4*s* electrons. The maximum oxidation state possible cannot exceed the total number of 4*s* and 3*d* electrons in the electron configuration.

- They have at least one oxidation state in which compounds and ions are coloured. The colours are often distinctive and can be used as a means of identification — for example, $Cu^{2+}(aq)$ ions are blue, $Cr^{3+}(aq)$ ions are green, $Cr_2O_7^{2-}(aq)$ ions are orange and $MnO_4^-(aq)$ ions are purple.
- Many transition metals are used as heterogeneous catalysts — for example, iron in the Haber process, nickel in the hydrogenation of alkenes, platinum, palladium and rhodium in catalytic converters. They work as catalysts because their *d*-orbitals bind other molecules or ions to their surfaces.

Simple precipitation reactions

Revised

A precipitation reaction takes place between an aqueous alkali and an aqueous solution of a metal(II) or metal(III) cation. This results in formation of a precipitate of the metal hydroxide, often with a characteristic colour. A suitable aqueous alkali is NaOH(aq). The colour of the precipitate can be used as a means of identification. Some precipitation reactions can be represented simply as follows:

$$Cu^{2+}(aq) + 2OH^-(aq) \rightarrow Cu(OH)_2(s)$$
Pale blue precipitate

$$Co^{2+}(aq) + 2OH^-(aq) \rightarrow Co(OH)_2(s)$$
Blue precipitate

$$Fe^{2+}(aq) + 2OH^-(aq) \rightarrow Fe(OH)_2(s)$$
Pale green gelatinous precipitate

$$Fe^{3+}(aq) + 3OH^-(aq) \rightarrow Fe(OH)_3(s)$$
Orange-brown gelatinous precipitate

$Fe(OH)_2(s)$ is slowly oxidised to $Fe(OH)_3(s)$. If left, the pale-green precipitate changes to an orange-brown precipitate.

Transition metal complexes

Revised

Ligands and complex ions

Transition metal ions are small and have a high charge density. They strongly attract electron-rich species called **ligands**, forming **complex ions**.

Common ligands include: H_2O**:**, **:**Cl^-, **:**NH_3, **:**CN^-, all of which have at least one lone pair of electrons.

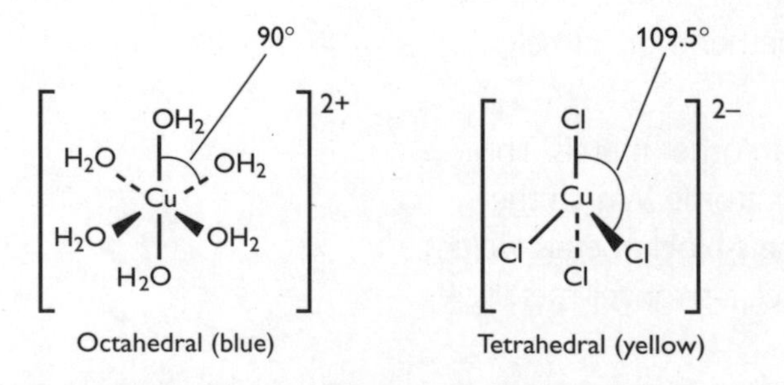

In $[Cu(H_2O)_6]^{2+}$, the six electron pairs surrounding the central Cu^{2+} ion repel one another as far apart as possible. The complex ion has an octahedral shape, so all the bond angles are 90°.

In $[CuCl_4]^{2-}$, the four electron pairs surrounding the central Cu^{2+} ion repel one another as far apart as possible. The complex ion has a tetrahedral shape, so all the bond angles are 109.5°.

A **ligand** is a molecule or ion that bonds to a metal ion forming a coordinate (dative covalent) bond by donating a lone pair of electrons into a vacant *d*-orbital.

A **complex ion** is defined as a central metal ion surrounded by ligands.

Coordination number

The **coordination number** of a transition metal indicates how many ligands there are around the metal ion.

Complex ions with ligands such as H_2O and NH_3 are usually 6-coordinate and octahedral in shape.

Complex ions with Cl^- ligands are usually 4-coordinate and tetrahedral in shape.

Ligands form a dative coordinate bond with a central transition metal ion. Some ligands are able to form two dative coordinate bonds with the central transition metal ion; these are known as **bidentate ligands**.

1,2-diaminoethane is a common bidentate ligand. Each nitrogen atom has a lone pair of electrons and each can form a dative bond.

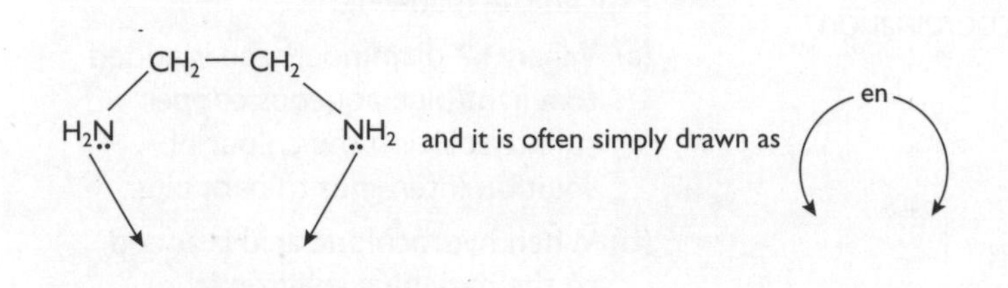

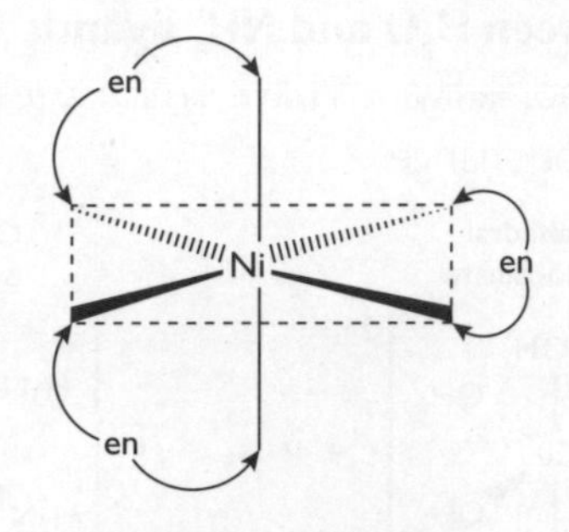

The **coordination number** is defined as the total number of coordinate bonds from the ligands to the central transition metal ion in a complex ion.

Stereoisomerism in complex ions

Isomerism is commonplace in organic compounds. It also occurs in some inorganic substances.

The square planar structure of $Ni(NH_3)_2Cl_2$ has two isomeric forms with the ammonia molecules and chloride ions being either on opposite sides of the complex ion (the *trans* form) or alongside each other (the *cis* form).

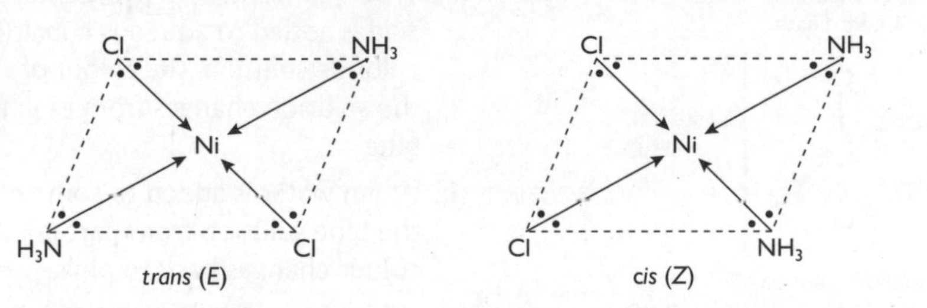

Examiner's tip

Any ion or molecule that has two nitrogens can act as a bidentate ligand and form complexes like the $Ni(en)_3^{2+}$ complex shown. To draw the complex all you have to do is replace 'en' with whatever you are given in the question. Diols and dicarboxylic acids can also behave as bidentate ligands because the oxygen atoms have lone pairs of electrons.

Complex ions have various uses. A particularly interesting example is the *cis* (*Z*) form of the molecule $PtCl_2(NH_3)_2$. (The platinum is present as platinum(II), so the complex therefore has no overall charge). The structure of $PtCl_2(NH_3)_2$ is:

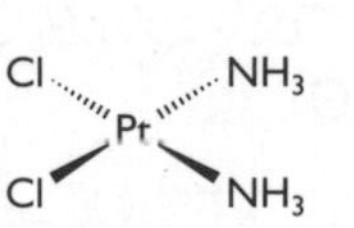

This compound is known as *cis*-platin. It is used during chemotherapy as an anti-cancer drug. It is a colourless liquid that is usually administered as a drip into a vein. It works by binding onto the DNA of cancerous cells and preventing their division. The importance of the exact shape and structure of the molecule is emphasised by the fact that the *trans* molecule is ineffective.

Optical isomerism is also possible in complexes coordinated with polydentate ligands. The nickel(II) 1,2-diaminoethane complex ion is an example:

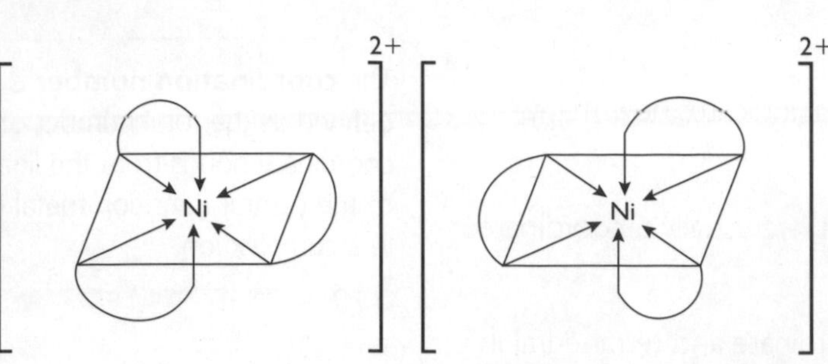

As is the case with organic molecules, it is the asymmetry of the structure that leads to this property. The two molecules shown cannot be superimposed on each other.

Ligand substitution of complex ions

A ligand substitution reaction takes place when a ligand in a complex ion exchanges with another ligand.

Exchange between H_2O and NH_3 ligands

Water and ammonia ligands have similar sizes, so the coordination number does not change.

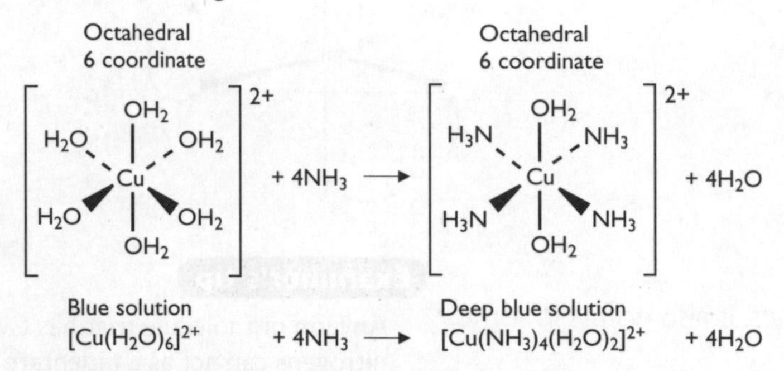

Exchange between H_2O and Cl^- ligands

Water molecules and chloride ions have different sizes, so the coordination number changes.

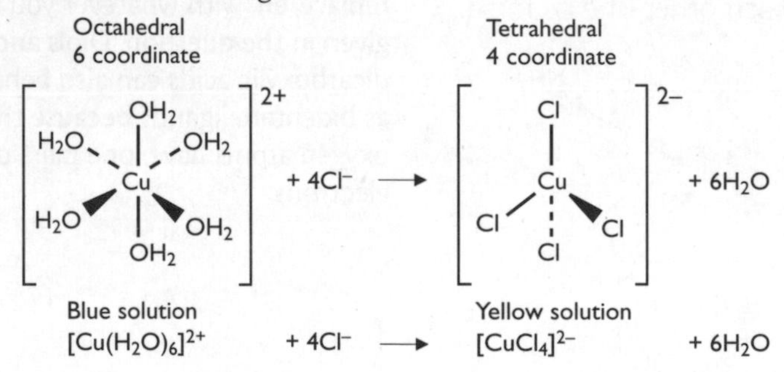

A similar reaction takes place when Co^{2+} replaces Cu^{2+}, in the formation of $[CoCl_4]^{2-}$ from $[Co(H_2O)_6]^{2+}$:

$$[Co(H_2O)_6]^{2+} + Cl^- \rightarrow [CoCl_4]^{2-} + 6H_2O$$

Pink Blue

Typical mistake

It is easy to lose marks by drawing the complex ion carelessly, showing the coordinate (dative) bond going from the wrong atom. In H_2O, the bond always goes from the oxygen, *not* the hydrogen; in NH_3 it always goes from the nitrogen.

Now test yourself

1 Explain the following:

(a) When 1,2-diaminoethane is added to a light blue aqueous copper sulfate solution the colour of solution intensifies to dark blue.

(b) When hydrochloric acid is added to the dark blue solution from part (a), the colour returns to the lighter blue.

Answers on p. 107

Now test yourself

2 Suggest explanations for each of the following:

(a) When concentrated hydrochloric acid is added to aqueous cobalt(II) chloride solution, the colour of the solution changes from pink to blue.

(b) When water is added to some of the blue solution from part (a), the colour changes back to pink.

(c) When aqueous silver nitrate is added to some of the blue solution from part (a) the solution changes to pink again and a precipitate is formed.

Answers on p. 107

Stability constants

Revised

The strength of binding of a ligand to a cation can be represented quantitatively.

When a complex ion such as $[Cu(NH_3)_4(H_2O)_2]^{2+}$ is formed it exists in equilibrium with the hydrated copper ion, $[Cu(H_2O)_6]^{2+}$ from which it was made. The equation is:

$$Cu(H_2O)_6^{2+} + 4NH_3 \rightleftharpoons Cu(NH_3)_4(H_2O)_2^{2+} + 4H_2O$$

As with other equilibria, this has an equilibrium constant. In this case, it is called the **stability constant**, K_{stab}:

$$K_{stab} = \frac{[Cu(NH_3)_4(H_2O)_2{}^{2+}]}{[Cu(H_2O)_6{}^{2+}][NH_3]^4} = 1.2 \times 10^{13}\ mol^{-4}\ dm^{12}$$

Free H_2O is not included in the expression.

The value of this equilibrium constant, $1.2 \times 10^{13}\ mol^{-4}\ dm^{12}$, indicates that the reaction lies well to the right and gives a measure of the greater stability of $[Cu(NH_3)_4(H_2O)_2]^{2+}$ compared with $[Cu(H_2O)_6]^{2+}$.

K_{stab} for $[CuCl_4]^{2-}$ is $4.2 \times 10^5\ mol^{-4}\ dm^{12}$. This is much smaller than the stability constant for the ammonia complex and reflects the fact that the $[CuCl_4]^{2-}$ complex is less stable than the $[Cu(NH_3)_4(H_2O)_2]^{2+}$ complex.

Redox reactions

Revised

Oxidation number

Oxidation number is a convenient way of identifying quickly whether a substance has undergone either oxidation or reduction. Many **redox** reactions occur in which the oxidation state of transition metal ions change through gaining or losing electrons. Oxidation and reduction can be identified by:

- movement of electrons — Oxidation Is Loss (of electrons) Reduction Is Gain (of electrons) (OILRIG)
- change in oxidation state/number — oxidation is an increase in oxidation number; reduction is a decrease.

The iron(II)–manganate(VII) reaction

The most common redox reaction involving transition elements that you will meet is the reaction between $Fe^{2+}(aq)$ and $MnO_4^-(aq)$.

Step 1: Write an ionic half-equation for each transition metal.

In this reaction, $Fe^{2+}(aq)$ is oxidised to $Fe^{3+}(aq)$. This can be written as a half-equation:

$$Fe^{2+} \rightarrow Fe^{3+} + 1e^- \qquad \text{(equation 1)}$$

Like any balanced equation, both the symbols and the charges have to balance.

If Fe^{2+} is oxidised, it follows that MnO_4^- must be reduced. It is reduced to Mn^{2+}. The first step in constructing a half-equation for this reduction is to recognise the change in oxidation state of the Mn.

As the oxidation number of oxygen is usually –2, it follows that in MnO_4^- the oxidation number of Mn is +7. The oxidation number of the Mn in Mn^{2+} is +2. Hence, the oxidation number of Mn changes from +7 to +2, so $5e^-$ must be gained:

$$MnO_4^- + 5e^- \rightarrow Mn^{2+}$$

This half-equation is *not* balanced. The reaction will not take place unless the MnO_4^- is acidified. Each oxygen in the MnO_4^- forms a water molecule. Since there are four oxygens in MnO_4^-, four water molecules will be formed requiring eight H^+:

$$MnO_4^- + 8H^+ + 5e^- \rightarrow Mn^{2+} + 4H_2O \qquad \text{(equation 2)}$$

Like any balanced equation, both the symbols and the charges have to balance — each side as a net charge of 2+.

Each half-equation is now balanced.

$Fe^{2+} \rightarrow Fe^{3+} + 1e^-$ equation (1)

$MnO_4^- + 8H^+ + 5e^- \rightarrow Mn^{2+} + 4H_2O$ equation (2)

Step 2: Rewrite the half-equations so that the number of electrons in both is the same.

In this case, we need to multiply equation 1 by 5 giving:

$5Fe^{2+} \rightarrow 5Fe^{3+} + 5e^-$

Equation 2 remains the same:

$MnO_4^- + 8H^+ + 5e^- \rightarrow Mn^{2+} + 4H_2O$

Step 3: Add the last two half-equations together to cancel out the electrons.

So the overall reaction equation is:

$5Fe^{2+} + MnO_4^- + 8H^+ \rightarrow 5Fe^{3+} + Mn^{2+} + 4H_2O$

(10+) + (1–) + (8+) = 17+
Net charge on the left-hand side

(15+) + (2+) = 17+
Net charge on the right-hand side

Like any balanced equation, the symbols and the charges have to balance.

Now test yourself

3 Deduce the oxidation number of the transition element in each of the following:

(a) $[Zn(NH_3)_4(H_2O)_2]^{2+}$
(b) $[Fe(CN)_6]^{4-}$
(c) $[Co(NH_3)_5Cl]^{2+}$
(d) $[Co(C_2O_4)_3]^{4-}$
(e) $[Cr(CH_3COO)_2(H_2O)_4]^+$

Answers on p. 107

Redox titrations

Transition metal ions are often coloured and the colour changes that occur when they react can be used to show when a titration has reached its end point. The reaction of Fe^{2+} with MnO_4^- is a good example. MnO_4^- is purple while Mn^{2+} is a very pale pink or almost colourless. When purple MnO_4^- is added from a burette into acidified Fe^{2+}, it immediately turns pale pink or colourless as the MnO_4^- reacts with the acidified Fe^{2+}. When all of the Fe^{2+} has reacted, the purple colour of further MnO_4^- added remains. The end point of this titration is when a faint permanent pink colour is seen.

The reaction between Fe^{2+} and MnO_4^- is often tested in the context of a titration calculation.

Example

Five iron tablets with a combined mass of 0.900 g were dissolved in acid and made up to 100 cm³ of solution. In a titration, 10.0 cm³ of this solution reacted exactly with 10.4 cm³ of 0.0100 mol dm⁻³ potassium manganate(VII). What is the percentage by mass of iron in the tablets?

Answer

Step 1: Write the balanced equation.

$5Fe^{2+}(aq) + MnO_4^-(aq) + 8H^+(aq) \rightarrow 5Fe^{3+}(aq) + Mn^{2+}(aq) + 4H_2O(l)$

Use the balanced equation to obtain the mole ratio of Fe^{2+} to MnO_4^-, which is 5:1.

Calculate the number of moles of MnO_4^- by using the concentration, c, and the reacting volume, *v*, of $KMnO_4$.

From the titration results, the amount of $KMnO_4$ can be calculated:

$$\text{amount of } KMnO_4 = c \times \frac{v}{1000} = 0.0100 \times \frac{10.4}{1000} = 1.04 \times 10^{-4}\text{ mol}$$

From the mole ratio, the amount of Fe^{2+} can be determined.

The Fe^{2+} to MnO_4^- ratio is 5:1. The amount of $KMnO_4$ is 1.04×10^{-4} mol.

Therefore, $5 \times 1.04 \times 10^{-4}$ mol Fe^{2+} reacts with 1.04×10^{-4} mol MnO_4^-

Amount of Fe^{2+} reacted = 5.20×10^{-4} mol

Step 2: Find the amount of Fe^{2+} in the solution prepared from the tablets.

10.0 cm^3 of $Fe^{2+}(aq)$ contains 5.20×10^{-4} mol $Fe^{2+}(aq)$

100 cm^3 solution of iron tablets contains $10 \times (5.20 \times 10^{-4}) = 5.20 \times 10^{-3}$ mol Fe^{2+}

Step 3: Find the percentage of Fe^{2+} in the tablets (A_r: Fe, 55.8).

5.20×10^{-3} mol Fe^{2+} has a mass of $5.20 \times 10^{-3} \times 55.8 = 0.290$ g

$$\% \text{ of } Fe^{2+} \text{ in tablets} = \frac{\text{mass of } Fe^{2+}}{\text{mass of tablets}} \times 100 = \frac{0.290}{0.900} \times 100 = 32.2\%$$

You also need to know the redox titration between iodine and thiosulfate ions, $S_2O_3^{2-}(aq)$:

$$2S_2O_3^{2-}(aq) + I_2(s) \rightarrow S_4O_6^{2-}(aq) + 2I^-(aq) \qquad \text{(equation 1)}$$

This titration is not usually used directly to determine the concentration of an iodine solution. Rather, it allows the determination of the concentration of a reagent that generates iodine as a result of a reaction.

An example is the determination of the concentration of a copper(II) sulfate solution. A known volume of copper(II) sulfate is reacted with excess potassium iodide:

$$2Cu^{2+}(aq) + 4I^-(aq) \rightarrow Cu_2I_2(s) + I_2(s) \qquad \text{(equation 2)}$$

Cu_2I_2 is copper(I) iodide, which forms as a grey-white precipitate. The reaction is, therefore, a redox process in which Cu^{2+} is reduced and I^- is oxidised.

The iodine produced is then titrated against a solution of sodium thiosulfate of known concentration.

At the start of the titration, the solution appears brown–purple because of the presence of the iodine. As the titration proceeds, this colour fades to yellow and the end point is reached when the solution is colourless. In practice, this colour change is quite difficult to see, mainly because of the presence of the copper(I) iodide precipitate. To help, some starch solution is added as the end point is approached. This gives rise to a dark-blue coloration that disappears sharply at the end point.

It can be seen that from:

equation (2) — 2 mol of Cu^{2+} react to produce 1 mol of I_2

equation (1) — 1 mol of I_2 reacts with 2 mol of $S_2O_3^{2-}$

It follows therefore that for every 1 mol of Cu^{2+}, 1 mol of $S_2O_3^{2-}$ is required. So the amount (in moles) of thiosulfate used is equivalent to the amount (in moles) of copper(II) ions present originally.

Check your understanding

1 Salts that contain a 'complexed' cation with an 'uncomplexed' anion can be crystallised. For example, cobalt forms a salt, **A**, of formula $[Co(NH_3)_6]Cl_3$ that is orange in colour.

Isomers of this cobalt complex can be made that retain the same number of chlorines but with a chloride ion taking the place of one of the ammonia molecules in the ligand. An example is salt, **B**, which has the formula $[Co(NH_3)_5(Cl)]Cl_2$.

(a) If equal volumes of solutions of the two salts **A** and **B** of the same concentration are reacted separately with excess aqueous silver nitrate, salt **A** produces a precipitate which has a mass that is 1.5 times greater than that produced by salt **B**. Explain why.

(b) Two different salts, **C** and **D**, can be made that contain Co^{3+} coordinated with four ammonia ligands. Each salt also contains three chlorines. When the experiment in (a) is repeated separately with **C** and **D**, each produces a mass of precipitate that is half the mass obtained from compound **B**.

Suggest structures for **C** and **D** and explain your answer.

2 A solution is made that contains the $VO^{2+}(aq)$ ion. When 25.0 cm³ of the solution is titrated against 0.0150 mol dm⁻³ $MnO_4^-(aq)$ ions in the presence of excess sulfuric acid, it is found that 23.3 cm³ are required to reach the end point. The equation for the oxidation of $VO^{2+}(aq)$ ions is:

$$VO^{2+}(aq) + 2H_2O(l) \rightarrow VO_3^-(aq) + 4H^+(aq) + e^-$$

Calculate the concentration in mol dm^{-3} of the solution containing $VO^{2+}(aq)$.

3 A general purpose solder contains antimony, lead and tin. When reacted with an acid, the solder dissolves to form a solution containing $Sb^{3+}(aq)$, $Pb^{2+}(aq)$ and $Sn^{2+}(aq)$. Neither $Sb^{3+}(aq)$ nor $Pb^{2+}(aq)$ reacts with dichromate ($Cr_2O_7^{2-}$) ions. $Sn^{2+}(aq)$ is oxidised to $Sn^{4+}(aq)$ by an acidified solution of potassium dichromate(VI).

In an experiment, 10.00 g of solder is dissolved in acid to make 1.00 dm³ of solution. When 25.0 cm³ of this solution is titrated against an acidified potassium dichromate solution of concentration 0.0175 mol dm^{-3}, 20.0 cm³ of the dichromate are required to reach an end point. The equation for the reduction of $Cr_2O_7^{2-}(aq)$ is:

$$Cr_2O_7^{2-}(aq) + 14H^+(aq) + 6e^- \rightarrow 2Cr^{3+}(aq) + 7H_2O(l)$$

(a) Write an overall equation for the reaction of $Sn^{2+}(aq)$ with $Cr_2O_7^{2-}(aq)$.

(b) Calculate the concentration of $Sn^{2+}(aq)$ in the solution in mol dm^{-3}.

(c) Calculate the percentage by mass of tin in the solder.

4 Rhubarb leaves contain poisonous ethanedioic acid, $(COOH)_2$. A dose of about 24 g of ethanedioic acid would probably be fatal if consumed by an adult. On heating, the ethanedioate ions react with potassium manganate(VII) solution and so this forms the basis for a titration. The ethanedioate ions are oxidised to carbon dioxide.

Four large rhubarb leaves are heated with water to extract the ethanedioic acid. The solution is filtered and the solution obtained is diluted to 250 cm³ in a volumetric flask.

25.0 cm³ samples of the solution are acidified with dilute sulfuric acid. 23.90 cm³ of 0.0200 mol dm^{-3} potassium manganate(VII) are required to reach the end point in a titration.

Calculate how many large rhubarb leaves would be needed to kill an adult.

5 A compound of formula $NaXO_3$ is produced from element X. An aqueous solution of $NaXO_3$ is made with a concentration of 0.0500 mol dm^{-3}. When excess potassium iodide solution is added to 25.0 cm³ of this solution, iodine is produced. When this is titrated against a solution of sodium thiosulfate containing 13.63 g dm^{-3}, 29.00 cm³ of the solution is required to react completely with the iodine.

Deduce the change in oxidation state of element X in the reaction.

Answers on pp. 108, 109

Exam practice

1 (a) (i) What is meant by the term: *transition element*? [2]

(ii) Copy and complete the electron configuration of the iron atom: $1s^22s^22p^6$................. [1]

(iii) Write the electron configuration of an Fe^{2+} ion and an Fe^{3+} ion. [2]

(b) (i) Aqueous Fe^{2+} ions react with aqueous hydroxide ions. Write an ionic equation for this reaction and state what you would see. [2]

(ii) The product formed in the reaction between aqueous Fe^{2+} ions and aqueous hydroxide ions slowly darkens and eventually turns 'rusty'. What has happened to cause this colour change? [2]

(c) The dichromate ion, $Cr_2O_7^{2-}$, is an oxidising agent that is used in laboratory analysis. It reacts with acidified Fe^{2+} ions to form Cr^{3+} and Fe^{3+} ions:

$$Cr_2O_7^{2-}(aq) + 14H^+(aq) + 6e^- \rightarrow 2Cr^{3+}(aq) + 7H_2O(l)$$

$$Fe^{2+}(aq) \rightarrow Fe^{3+}(aq) + e^-$$

(i) Construct the full ionic equation for this reaction. [1]

(ii) Calculate the volume of 0.0100 mol dm^{-3} potassium dichromate required to react with 20.0 cm^3 of 0.0500 mol dm^{-3} acidified iron(II) sulfate. [3]

2 (a) Explain the meaning of the terms: *ligand* and *coordinate bond.* [2]

(b) Stereoisomerism is sometimes shown by transition metal complex ions. Using a suitable named example in each case, show how transition metal complex ions can form:

(i) *cis–trans* isomers

(ii) optical isomers [8]

(c) Ligand exchange may occur when transition metal complex ions react. This will take place if a new complex ion that has a greater stability can be formed. An example of a ligand exchange reaction is the formation of $[CoCl_4]^{2-}$ from $[Co(H_2O)_6]^{2+}$.

(i) The stability of a complex ion is expressed by its stability constant. Define the stability constant of $[CoCl_4]^{2-}$. [2]

(ii) Describe how you would convert $[Co(H_2O)_6]^{2+}$ into $[CoCl_4]^{2-}$. What you would see as the reaction takes place? [3]

3 Synoptic Question

Most questions on this paper are based on the specification content, but you are likely to be given data and asked to use those data to answer the question. It might involve extended writing or a more lengthy calculation. The question that follows is along these lines.

Copper reacts with nitric acid, HNO_3, but the products of the reaction depend on the concentration of the acid.

If the nitric acid is dilute, the following reaction takes place:

$$Cu(s) \rightarrow Cu^{2+}(aq) + 2e^-$$

$$4H^+(aq) + NO_3^-(aq) + 3e^- \rightarrow NO(g) + 2H_2O(l)$$

If the nitric acid is concentrated, the reaction is:

$$Cu(s) \rightarrow Cu^{2+}(aq) + 2e^-$$

$$2H^+(aq) + NO_3^-(aq) + e^- \rightarrow NO_2(g) + H_2O(l)$$

(a) Write balanced equations for each of the above reactions. [5]

(b) In an experiment, some nitric acid is reacted with 1.27 g of copper. It is found that 320 cm^3 of gas is produced. Deduce whether the acid used in this experiment was dilute or concentrated. Show *all* your working. [4]

Answers and quick quiz 6 online

Online

Examiner's summary

You should now have an understanding of:

- ✔ general properties of transition elements
- ✔ precipitation reactions
- ✔ ligands and complex ions
- ✔ ligand substitution reactions
- ✔ redox reactions and titrations

Answers

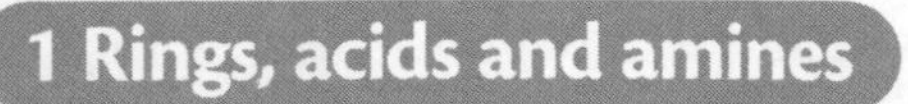

1 Rings, acids and amines

Now test yourself

1

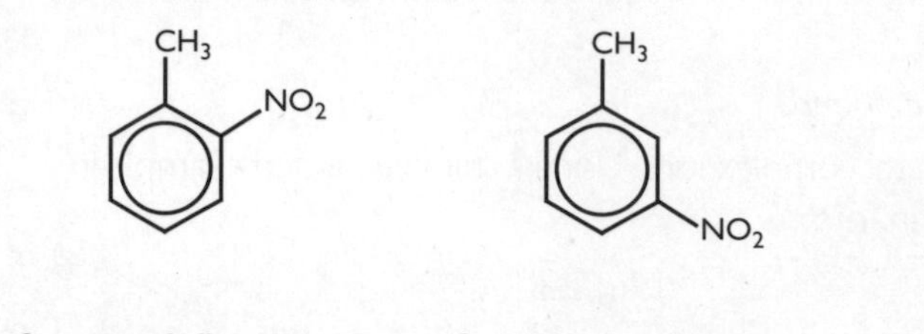

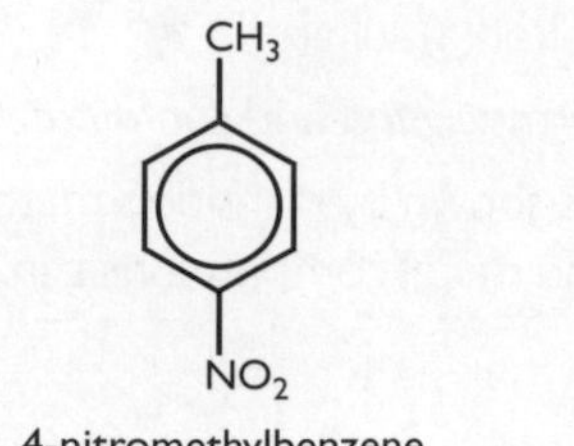

2-nitromethylbenzene 3-nitromethylbenzene 4-nitromethylbenzene

2 $C_6H_5CH_3 + 3HNO_3 \rightarrow CH_3C_6H_2(NO_2)_3 + 3H_2O$

3 C_7H_8O

4 (a) **A** = 2-methylpropanal

B = 3-methylbutan-2-one

C = phenylethanal

D = phenylethanone

E = 2,2-dimethylpropanal

F = cyclohexane-1,4-dione

(b) A, C and E are aldehydes; B,D and F are ketones

(c) $C_6H_8O_2$

(d) C could be prepared from 2-phenylethanol; D could be prepared from 1-phenylethanol

(e) Any three from:

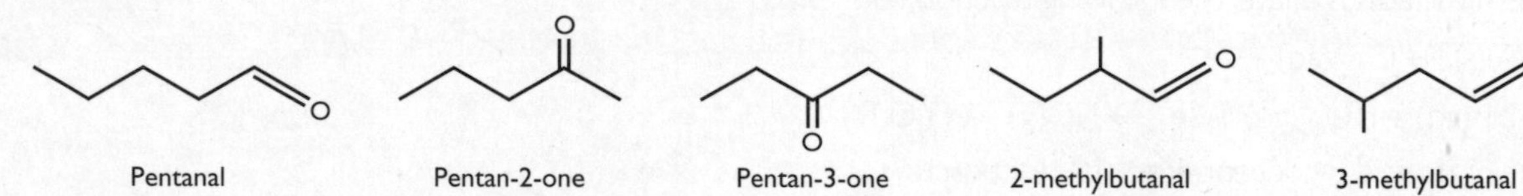

Pentanal Pentan-2-one Pentan-3-one 2-methylbutanal 3-methylbutanal

5 (a) Reaction with Na(s):

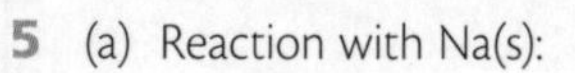

$HCOOH(aq) + Na(s) \rightarrow HCOONa(aq) + ½\ H_2(g)$

$H{-}C({=}O){-}O{-}H + Na \longrightarrow H{-}C({=}O){-}O^-Na^+ + ½H_2$

(b) Reaction with $NaHCO_3(aq)$:

$H{-}C({=}O){-}O{-}H + NaHCO_3 \longrightarrow H{-}C({=}O){-}O^-Na^+ + H_2O + CO_2$

$HCOOH(aq) + NaHCO_3(aq) \rightarrow HCOONa(aq) + H_2O(l) + CO_2(g)$

6 (a) Refluxing is continuous evaporation and condensation on heating so that volatile components do not escape.

(b)

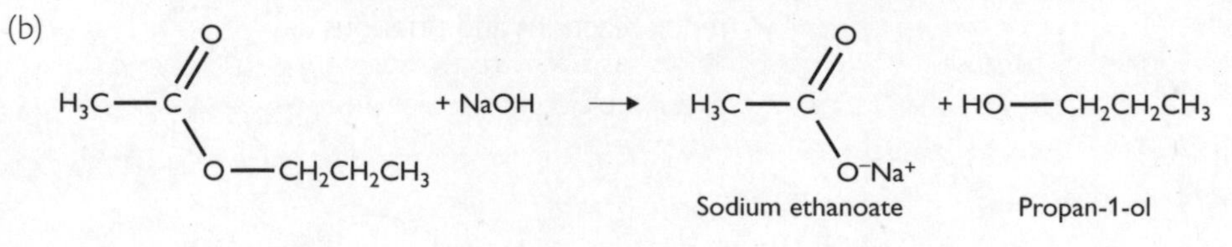

Sodium ethanoate Propan-1-ol

7 *Cis* (*Z*) fatty acids have a distinctive 'kinked' shape.
Trans (*E*) fatty acids, the kink is straightened out.

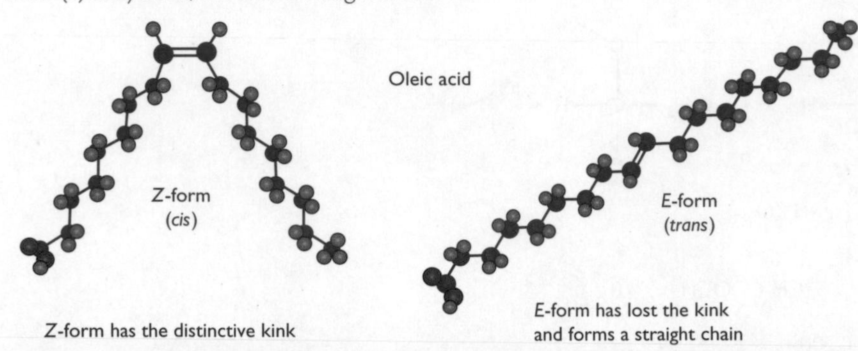

One possible disadvantage of a diet rich in *trans* fatty acids is that it might lead to a build-up of plaques in the arteries, which could result in heart disease.

8

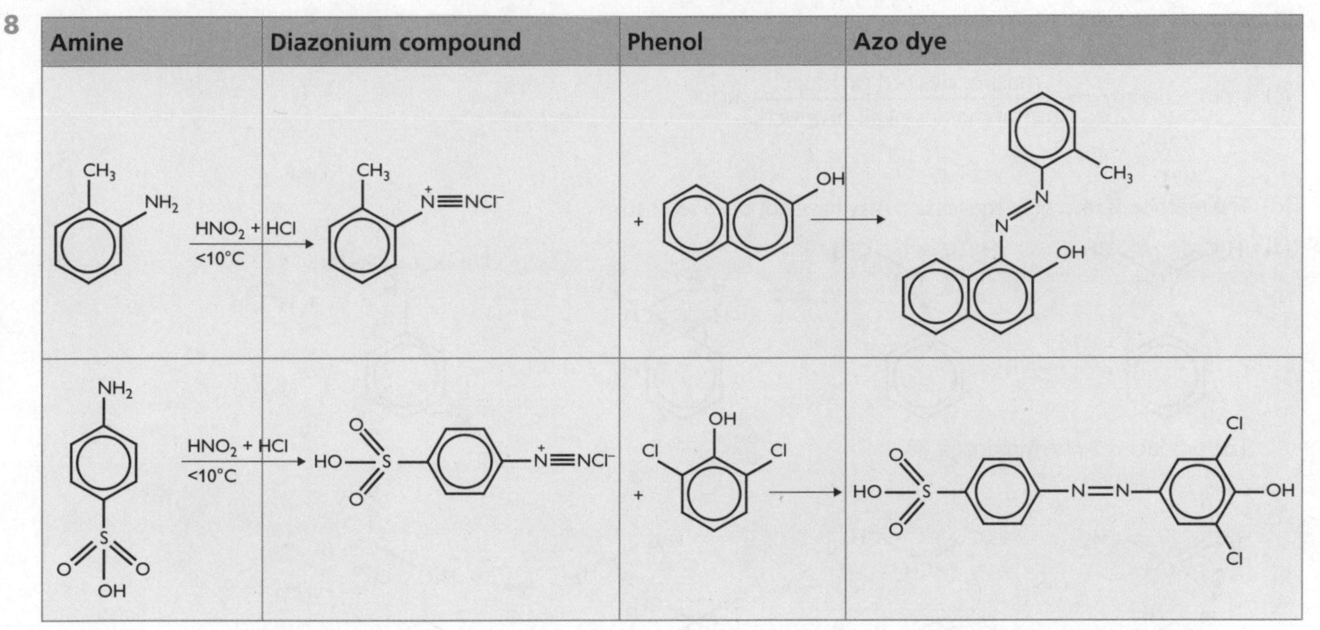

Check your understanding

1 (a)

	Benzene	Phenol	Cyclohexene
Type of reaction	Electrophilic substitution	Electrophilic substitution	Electrophilic addition
Reagents and conditions (if any)	Br_2 + halogen carrier, such as $AlBr_3$ or Fe	Br_2 No special conditions	
Organic product and observations	Bromobenzene, C_6H_5Br Br_2 is decolorised	2,4,6-tribromophenol White precipitate; Br_2 is decolorised	1,2-dibromocyclohexane Br_2 is decolorised

(b) (i) **Benzene and phenol**

Bromine reacts faster with phenol because one of the lone pairs of electrons on the oxygen in the OH group is delocalised into the ring. This increases the electron density which in turn polarises the Br–Br bond, so that an electrophile is generated. This then attacks the ring. (Accept the reverse argument for why the reaction with benzene is slower.)

(ii) **Benzene and cyclohexene**

Bromine reacts faster with cyclohexene because the C=C double bond has a high electron density which polarises the Br–Br bond, so that an electrophile is generated. This then attacks the C=C double bond. (Accept the reverse argument for why the reaction with benzene is slower)

2 Carbonyl compound used: butanone, $CH_3COCH_2CH_3$
Balanced equation: $CH_3COCH_2CH_3 + 2[H] \rightarrow CH_3CH(OH)CH_2CH_3$
Reagents and conditions: $NaBH_4$; water as solvent

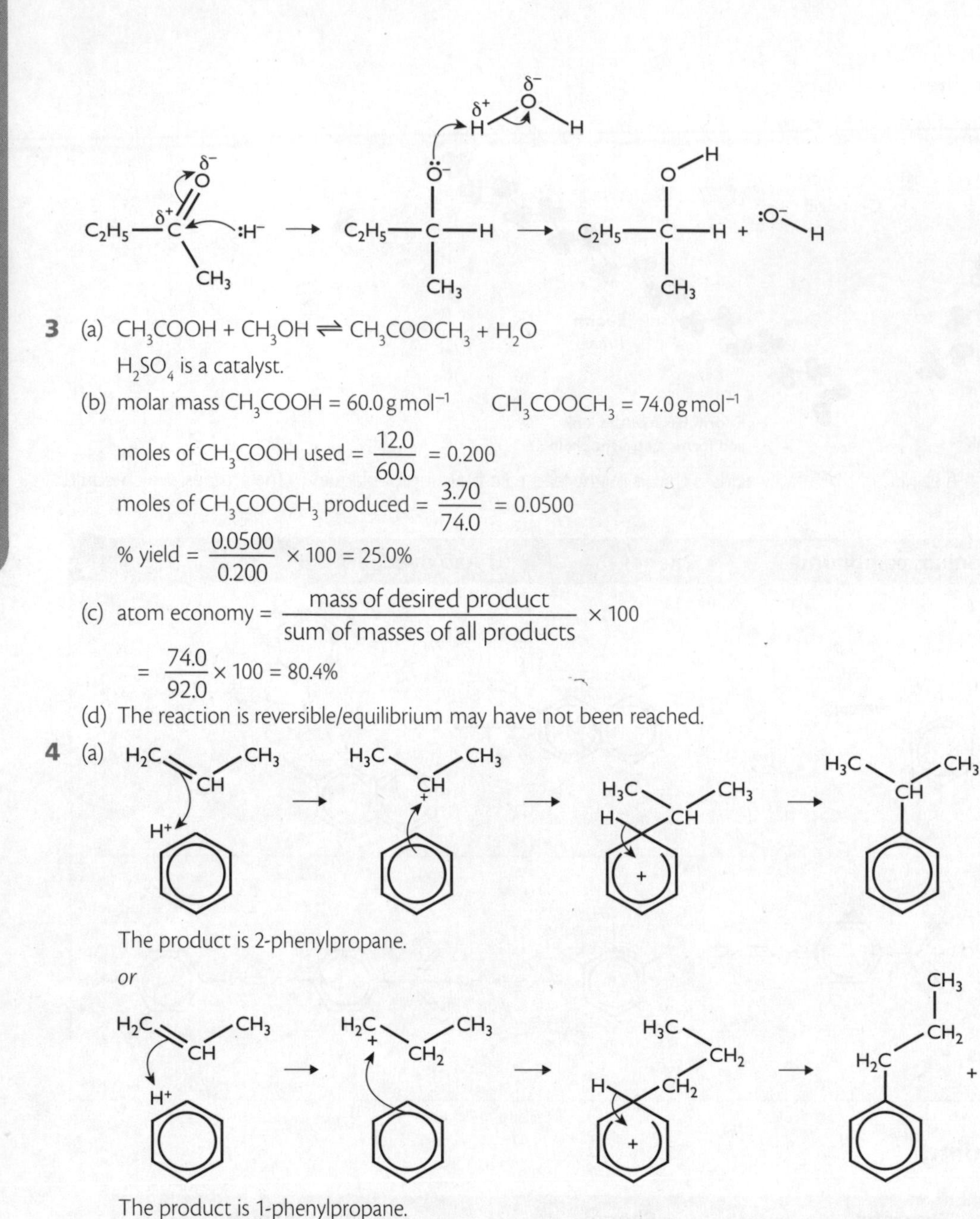

3 (a) $CH_3COOH + CH_3OH \rightleftharpoons CH_3COOCH_3 + H_2O$

H_2SO_4 is a catalyst.

(b) molar mass $CH_3COOH = 60.0\,g\,mol^{-1}$ $\quad CH_3COOCH_3 = 74.0\,g\,mol^{-1}$

moles of CH_3COOH used $= \frac{12.0}{60.0} = 0.200$

moles of CH_3COOCH_3 produced $= \frac{3.70}{74.0} = 0.0500$

% yield $= \frac{0.0500}{0.200} \times 100 = 25.0\%$

(c) atom economy $= \frac{\text{mass of desired product}}{\text{sum of masses of all products}} \times 100$

$= \frac{74.0}{92.0} \times 100 = 80.4\%$

(d) The reaction is reversible/equilibrium may have not been reached.

4 (a)

The product is 2-phenylpropane.

or

The product is 1-phenylpropane.

(b) The two possible products are 2-phenylpropane and 1-phenylpropane, as shown above. The product obtained depends on which carbonium ion (carbocation) is formed initially.

(c)

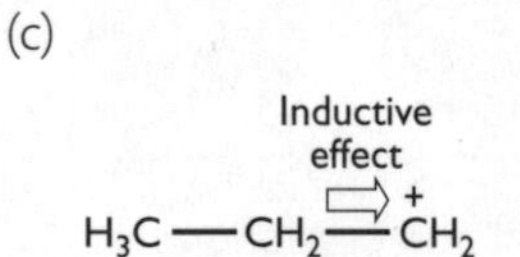

Primary carbonium ion

$H_3C—\overset{+}{\underset{|}{C}}(H)—CH_3$

Inductive effects

Secondary carbonium ion

Carbonium ions are unstable. They may not exist long enough to collide and react. Anything that stabilises the carbonium ion increases the chance of it existing long enough for collision and reaction to occur.

Carbonium ions are stabilised by the inductive effect of adjacent alkyl groups, which release electrons along the σ-bond. The primary carbonium ion is stabilised by the adjacent ethyl group, C_2H_5; the secondary carbonium ion is stabilised by both adjacent methyl groups. This makes the secondary carbonium ion more stable and, therefore, more likely to exist long enough to collide and react. Hence the major product is 2-phenylpropane.

5 (a) When ammonia, NH_3, reacts with chloroalkanes it behaves as a nucleophile because there is a lone pair of electrons on the nitrogen. Ammonia reacts with water producing NH_4^+ and OH^- ions. When NH_4^+ is formed, the nitrogen loses its lone pair of electrons so it can no longer behave as a nucleophile.

(b) NH_4OH is a weak base and exists in the equilibrium:

$$NH_3(g) + H_2O(l) \rightleftharpoons NH_4^+(aq) + OH^-(aq)$$

When $NH_4Cl(aq)$ is added to aqueous ammonia there is a large excess of $NH_4^+(aq)$ ions. This forces the equilibrium to the left, enabling ammonia to exist as $:NH_3$, so it can behave as a nucleophile.

(c) **Step 1**

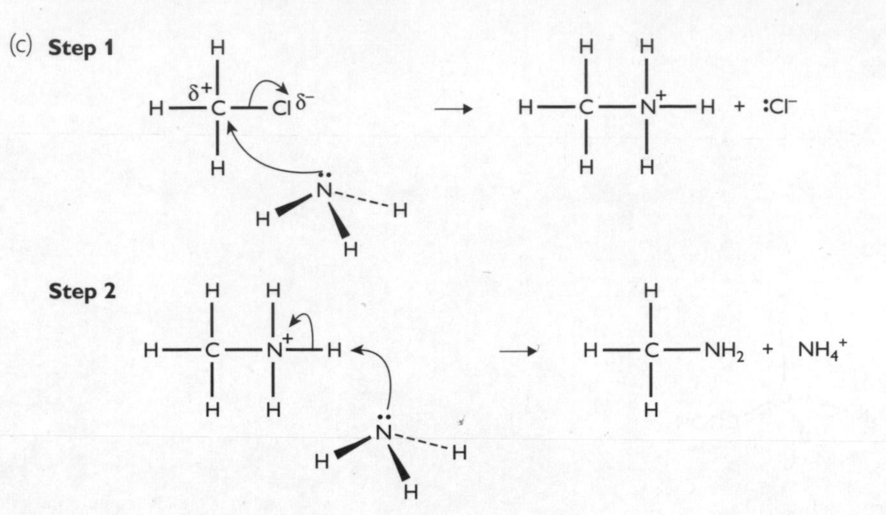

(d) (i) To produce $CH_3CH_2NH_2$, use a large excess of NH_3.

(ii) To obtain $(CH_3CH_2)_4N^+$, use a large excess of chloroethane.

6

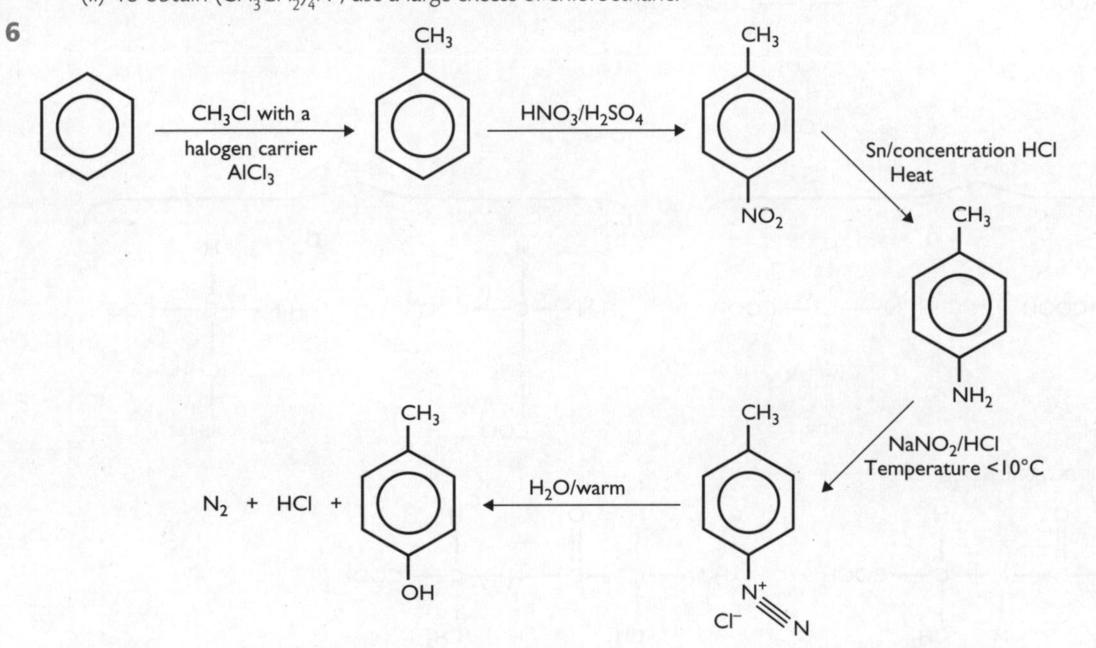

This reaction involves five separate steps and each step will introduce impurities. For example, in the second step, the nitration of methylbenzene, it is likely that the 2- and 6-positions will also be nitrated and that a mixture of mono-, di- and tri-nitrated products will be produced.

7

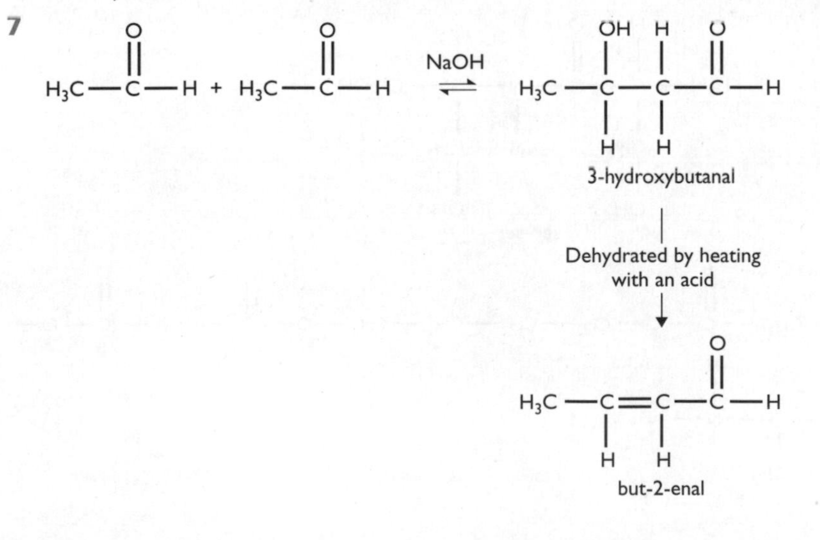

2 Polymers and synthesis

Now test yourself

1 (a) A 2-aminopropanoic acid

B 2-aminobutane-1,4-dioic acid

C 2-amino-3-hydroxypropanoic acid

D 2,6-diaminohexanoic acid

(b)

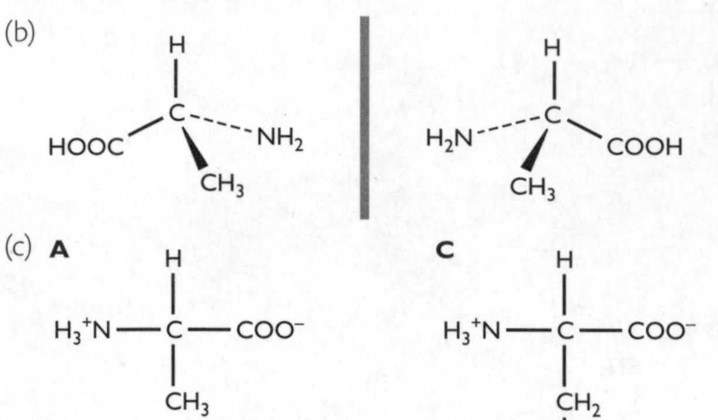

(c)

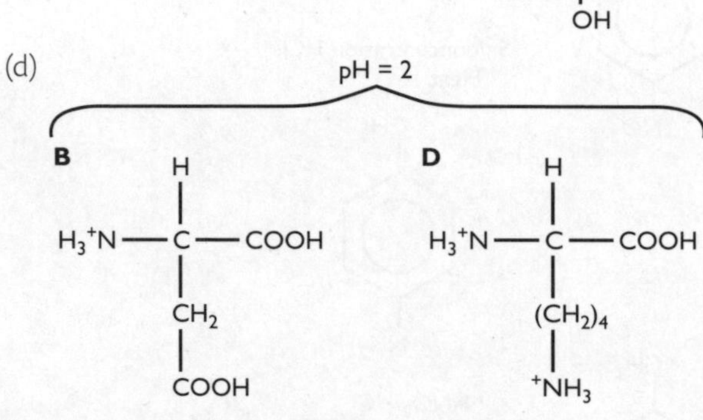

(d)

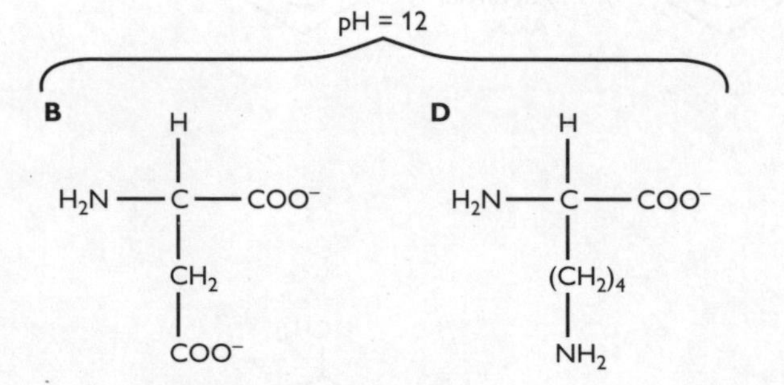

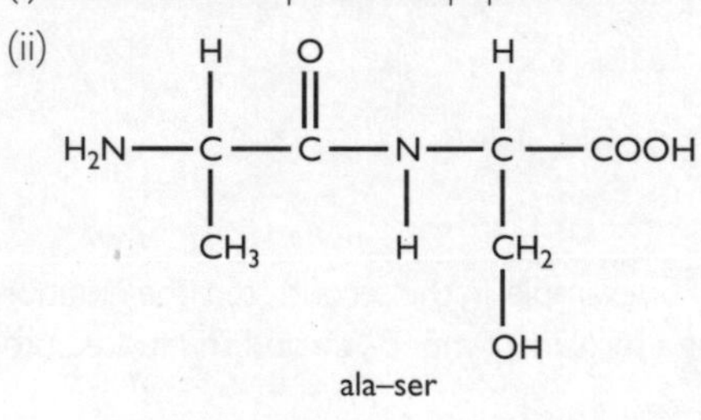

(e) (i) There are four possible dipeptides.

(ii)

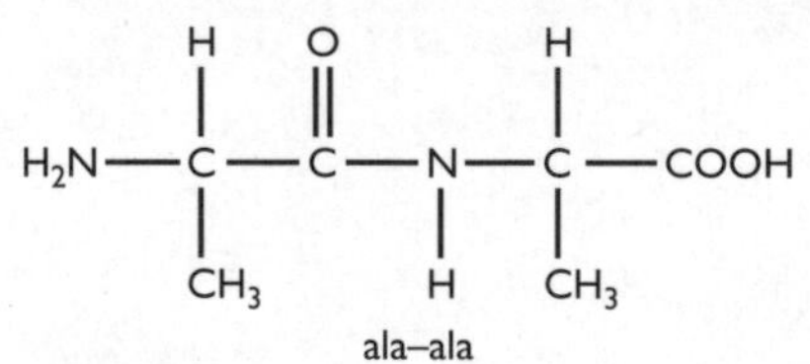

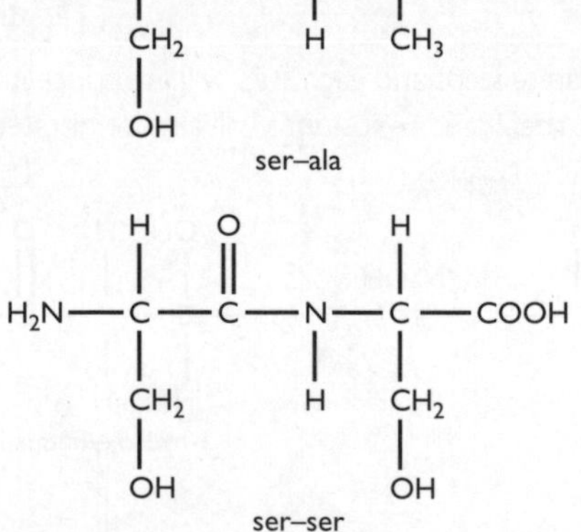

2 (a)

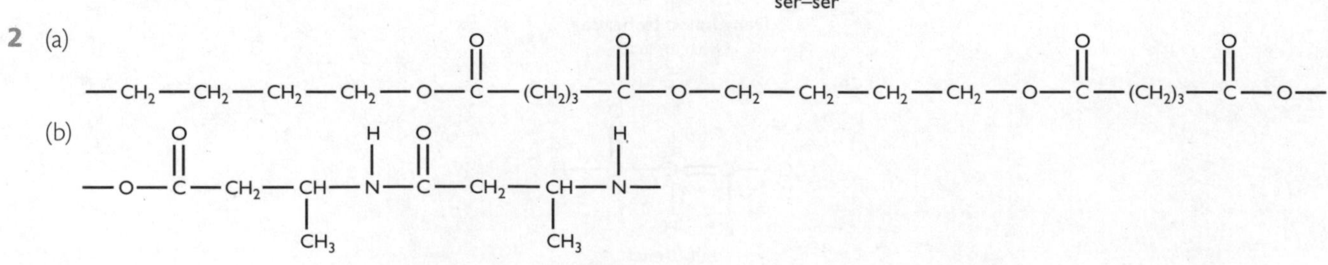

(b)

3 Acid hydrolysis gives:

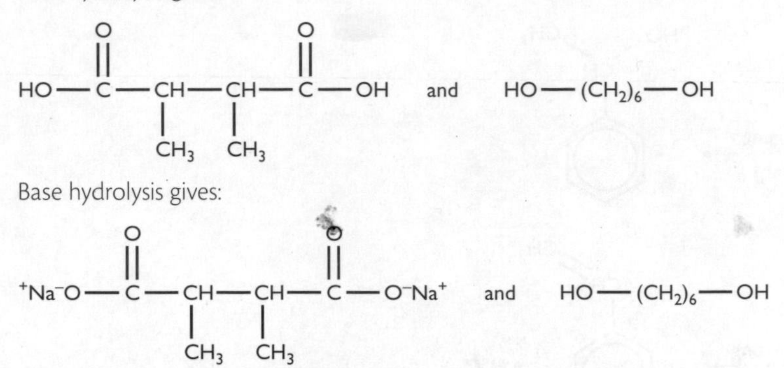

Check your understanding

1 (a) (i) **nylon-4, 6** 1,4-diaminobutane and hexane-1,6-dioic acid (or hexane-1,6-dioyl chloride)
nylon-6, 4 1,6-diaminohexane and butane-1,4-dioic acid (or butane-1,4-dioyl chloride)
nylon-6,10 1,6-diaminohexane and decane-1,10-dioic acid (or decane-1,10-dioyl chloride)

(ii) Melting point depends on the strength and amount of intermolecular forces. Nylon-4,6 and nylon-6,10 have an amide link that enables them to form hydrogen bonds with adjacent strands of nylon. There are also van der Waals forces. Nylon-6,10 has more electrons than nylon-4,6, so nylon-6,10 has more van der Waals forces and, therefore, has a higher melting point.

(b) (i)

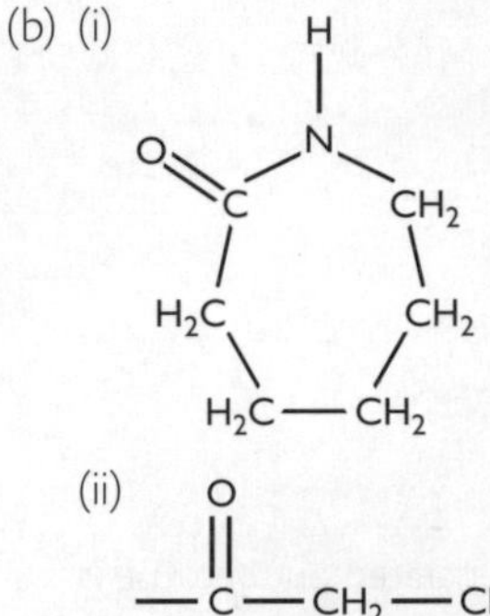

(ii)

$$-C(=O)-CH_2-CH_2-CH_2-CH_2-CH_2-N(H)-C(=O)-CH_2-CH_2-CH_2-CH_2-CH_2-N(H)-$$

2 (a) 1 mark for each step.

The simplest way is:

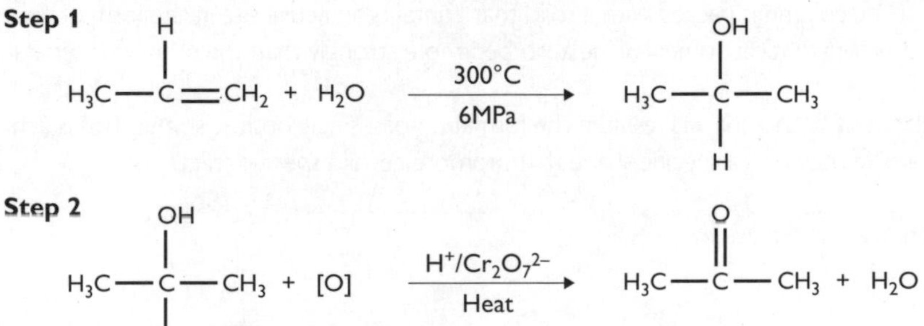

(b) 1 mark for each step.

The simplest way is:

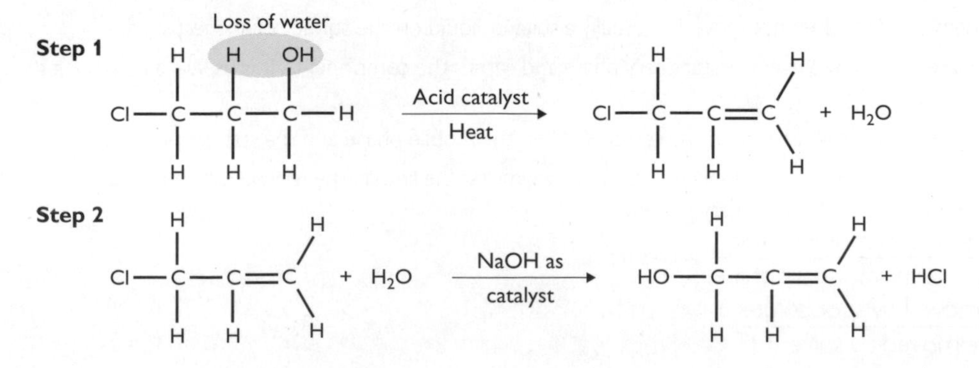

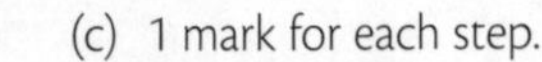

(c) 1 mark for each step.

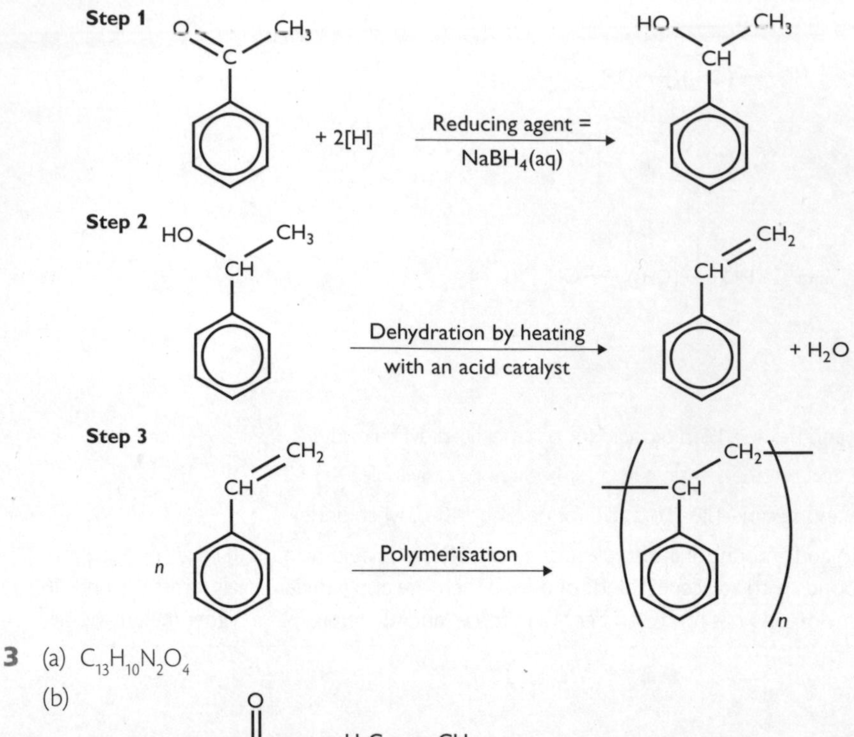

3 (a) $C_{13}H_{10}N_2O_4$

(b)

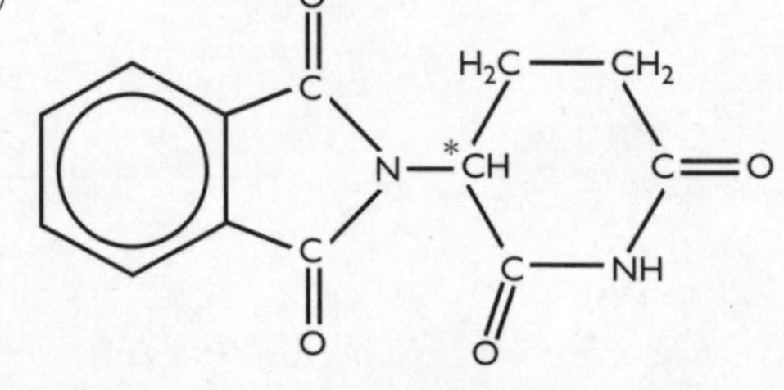

(c) Only one of the optical isomers has the correct shape to be active pharmaceutically. It has to be separated out because it is possible that the other isomer could have adverse side effects or make the correct isomer less effective. Therefore, it is vital that the effective isomer is isolated.

(d) Any two from the following:

- Chiral chromatography can be used. If the column is packed with a solid that contains an active site in the form of either an enzyme or a chiral stationary phase, then one optical isomer will be absorbed more strongly than the other and separation will be achieved.
- Synthesis using naturally occurring enzymes or bacteria will result in the formation of a single optical isomer. This is achieved because enzymes and bacteria have active sites with a specific shape that promote certain specific reactions.
- chiral catalysis
- chiral synthesis using naturally occurring chiral molecules

3 Analysis

Now test yourself

1 (a) **stationary phase** — the phase that does not move. It is usually a solid or liquid on the surface of an inert solid.

(b) **mobile phase** — the phase that moves over the stationary phase and carries the components. It is usually a liquid or a gas/solvent.

(c) **adsorption** — the interaction/attraction between the components in the mobile phase and the stationary phase.

(d) **partition** — the ratio of the solubility of a component between two immiscible liquids/the relative distribution of the component between the mobile and the stationary phases.

$$\frac{\text{concentration of solute in mobile phase}}{\text{concentration of solute in stationary phase}} = \text{constant}$$

(e) $\mathbf{R_f}$ **value** $= \dfrac{\text{distance moved by spot/solute}}{\text{distance moved by solvent}}$

(f) **retention time** — the time from the injection of the sample for a component to leave the column.

2

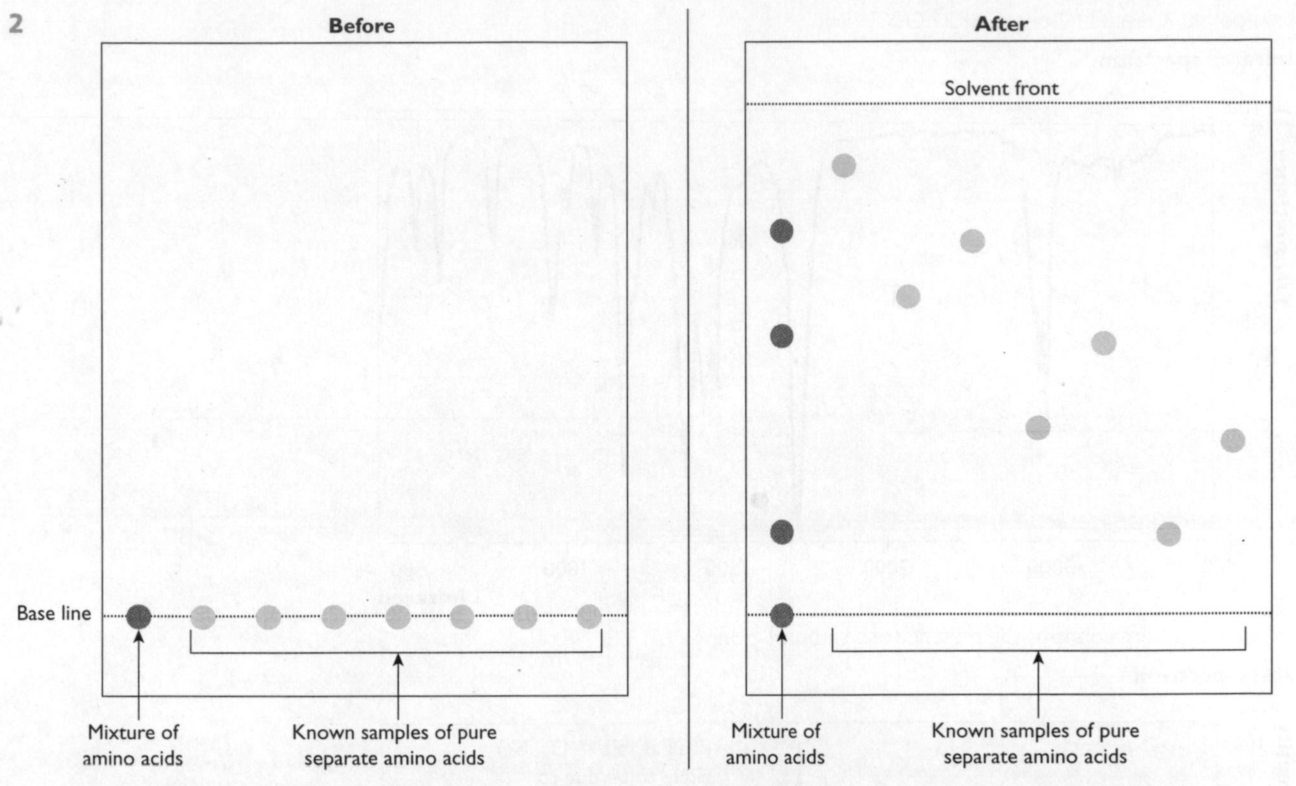

The mixture of amino acids is spotted onto the base line and known samples of pure amino acids are spotted alongside. The plate is then placed in a suitable solvent and left. When the solvent has moved almost to the end of the plate, the plate is removed and the position of the solvent front is recorded. The plate is dried and developed using ninhydrin. The positions of the spots in the mixture are compared with the position of the spots in the reference samples.

3 Any four from the following:

- Forensics — GC–MS can be used to analyse particles (perhaps from a human body) in order to help link a criminal to a crime, and to detect and identify small amounts of narcotics
- Airport security — detection of explosives
- Food and drink analysis — identification of compounds such as esters, fatty acids, and alcohols. It is also used to measure contamination by pesticides and herbicides.
- Medicine — analysis of pharmaceuticals and of metabolic compounds labelled with ^{13}C
- Astrochemistry — obtaining information about other planets, including Mars and Saturn

Check your understanding

1 (a) ^{1}H NMR detects the two proton environments labelled H_a and H_b in the diagram on the left; ^{13}C NMR detects the three carbon environments labelled C_1, C_2 and C_3 on the right.

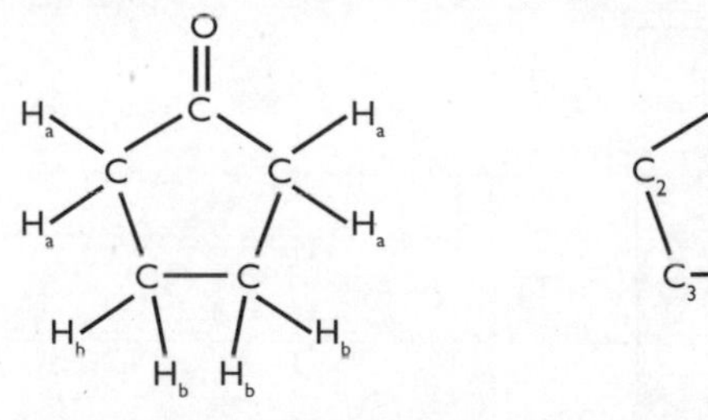

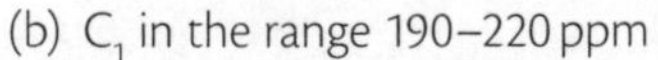

(b) C_1 in the range 190–220 ppm

C_2 in the range 20–30 ppm (5–55 ppm allowed)

C_3 in the range 5–55 ppm

(c) Chemical shifts: H_a in the range 2.0–2.9 ppm; H_b in the range 0.7–2.0 ppm

Splitting patterns: H_a and H_b would both be split into a triplet as in each case the adjacent carbon has two hydrogens

Relative peak area: 1 : 1 as there are four hydrogens in environment H_a and four hydrogens in environment H_b

2 Compound **X** is butan-2-one, $CH_3CH_2COCH_3$.

Infrared spectrum

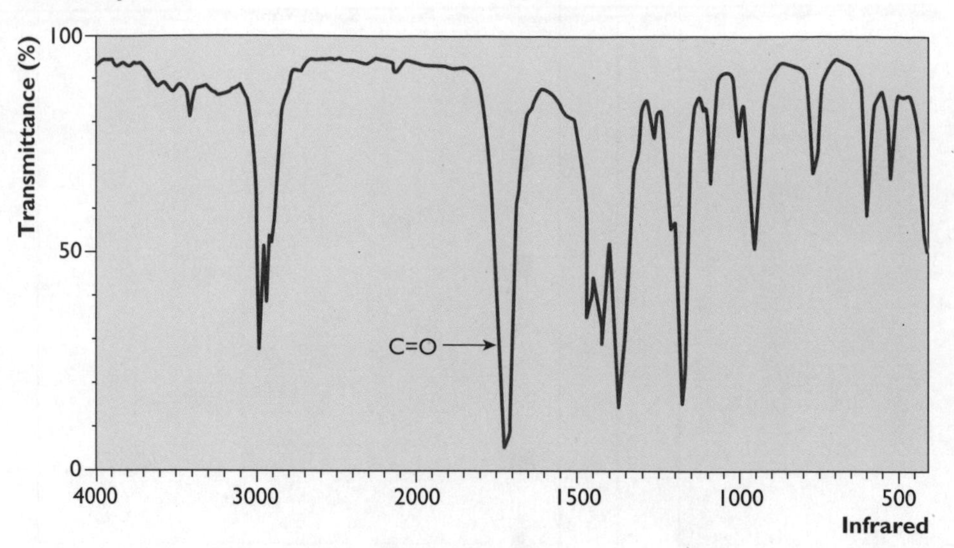

Infrared spectrum confirms the presence of a carbonyl group.

Mass spectrum

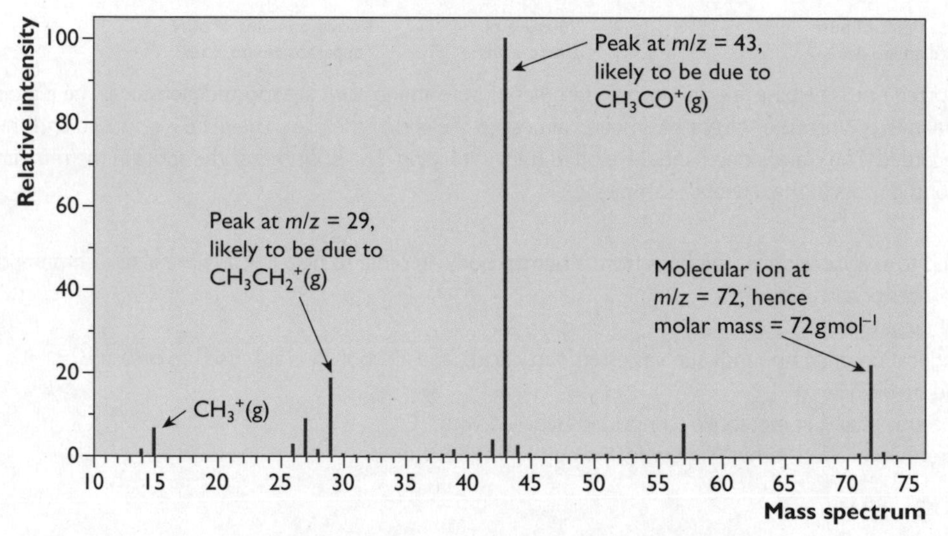

Mass spectrum indicates the presence of two fragments: $CH_3CO^+(g)$ and $CH_3CH_2^+(g)$. These add up to the molecular ion.

^{13}C-NMR

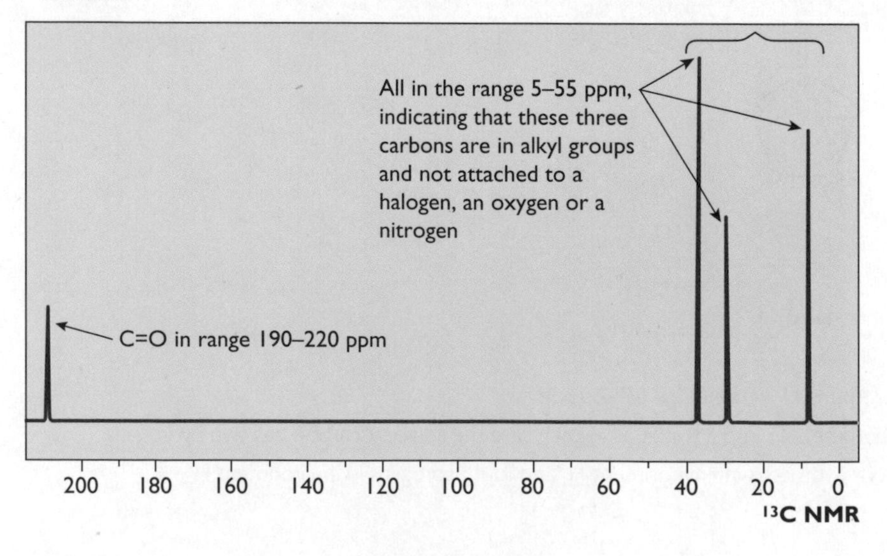

^{13}C-NMR confirms the presence of four different carbon environments, one of which is a C=O.

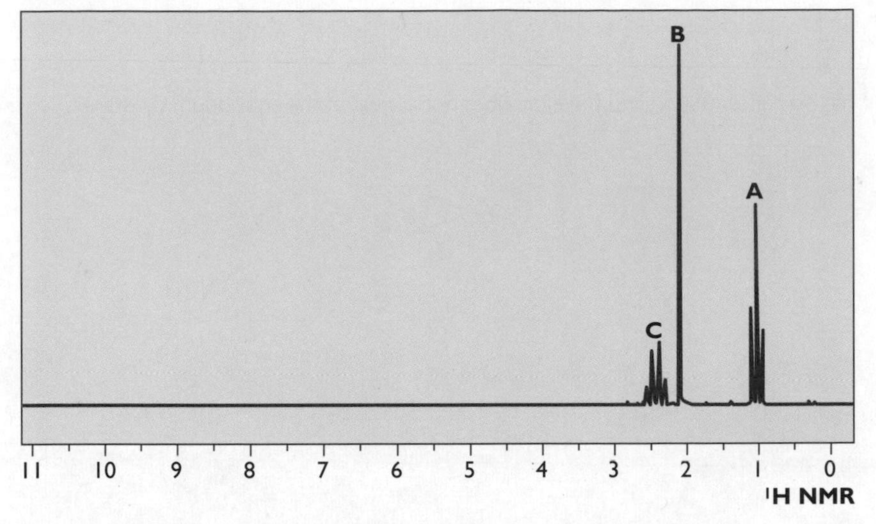

^{1}H-NMR confirms the presence of three different hydrogen environments.

Peaks **A** and **C** show classic triplet and quartet splitting confirming the presence of a CH_3 group next to a CH_2 group. This also confirms that the carbon on the other side of the CH_2 has no hydrogens attached. Peak **B** is a singlet, which confirms that it is bonded to a carbon that has no hydrogens attached.

3

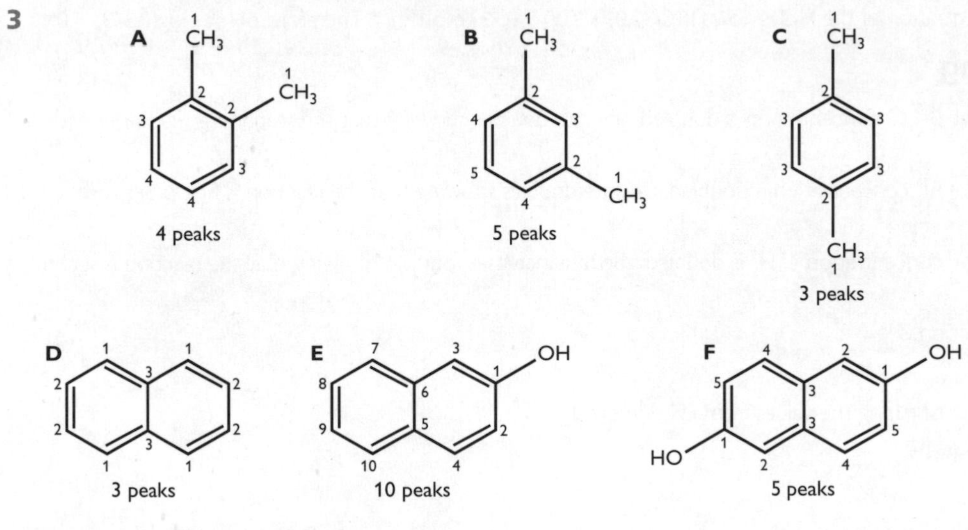

4 Rates, equilibria and pH

Now test yourself

1 (a) Fourth order; units of k = mol^{-3} dm^{12} s^{-1} (or dm^{12} mol^{-3} s^{-1})

(b) (i) The rate is three times as fast.

(ii) The rate is $(\frac{1}{2})^2 = \frac{1}{4}$

(iii) The rate increases by $2 \times 2^2 \times 2 = 16$

2 (a) $K_c = \frac{[NO_2(g)]^2}{[N_2O_4(g)]}$

Units: mol dm^{-3}

(b) $K_c = \frac{[N_2(g)]^2[O_2(g)]}{[N_2O(g)]^2}$

Units: mol dm^{-3}

(c) $K_c = \frac{[CO(g)]^2}{[O_2(g)]}$

Units: $mol\,dm^{-3}$

Remember that as carbon is a solid, it has no concentration and therefore does not appear in the equilibrium constant expression.

(d) $K_c = \frac{[CO_2(g)]^2}{[CO(g)]^2[O_2(g)]}$

Units: $mol^{-1}\,dm^3$ or $dm^3\,mol^{-1}$

3 (a) $HCO_3^- + H_2O \rightleftharpoons H_2CO_3 + OH^-$

Base 2   Acid 1   Acid 2   Base 1

(b) $HCO_3^- + OH^- \rightleftharpoons H_2O + CO_3^{2-}$

Acid 1   Base 2   Acid 2   Base 1

4 (a) pH = −log(0.15) = 0.82

(b) pH = $-\log(7.9 \times 10^{-6}) = 5.1$

(c) pH = $-\log(6.67 \times 10^{-14}) = 13.2$

5 When $20.0\,cm^3$ of $1.00\,mol\,dm^{-3}$ HCl is added to $10.0\,cm^3$ of $1.00\,mol\,dm^{-3}$ NaOH a reaction occurs and the NaOH is neutralised, producing neutral NaCl.

$10.0\,cm^3$ of the $1.00\,mol\,dm^{-3}$, HCl remains un-neutralised. However, this has been diluted to the total volume of the mixture, which is $30.0\,cm^3$. This means the concentration of the HCl is now $(10.0/30.0) \times 1.00 = 0.333\,mol\,dm^{-3}$. Therefore, pH = −log(0.333) = 0.48.

Check your understanding

1 (a) Experiments 1 and 3: when the BrO_3^- concentration is doubled, the rate doubles showing that the reaction is first order with respect to BrO_3^-.

Experiments 2 and 3: when the Br^- concentration is doubled, the rate doubles showing that the reaction is first order with respect to Br^-.

Experiments 2 and 4: when the concentration of H^+ is doubled, the rate increases fourfold, showing that the reaction is second order with respect to H^+.

The rate equation for the reaction is:

rate = $k[BrO_3^-][Br^-][H^+]^2$

(b) Using the rate equation and substituting the values from Experiment 1:

$1.64 \times 10^{-3} = k \times 0.1 \times 0.2 \times (0.1)^2$

$$k = \frac{1.64 \times 10^{-3}}{0.1 \times 0.2 \times (0.1)^2}$$

$$= \frac{1.64 \times 10^{-3}}{2 \times 10^{-4}} = 8.2\,dm^9\,mol^{-3}\,s^{-1}$$

(c) rate = $8.2 \times 0.25 \times 0.25 \times (0.25)^2 = 3.2 \times 10^{-2}\,mol\,dm^{-3}\,s^{-1}$

2 (a) It is helpful to tabulate initial and equilibrium amounts:

	Units	N_2O	N_2	O_2
Initial moles	mol	1.00	0	0
Equilibrium moles	mol	0.10	0.90	0.45
Equilibrium concentrations	$mol\,dm^{-3}$	0.10	0.90	0.45

(b) Remember: when calculating K_c, you *must* use concentrations, *not* moles.

$$K_c = \frac{[N_2(g)]^2[O_2(g)]}{[N_2O(g)]^2}$$

$$= \frac{(0.90)^2 \times 0.45}{(0.10)^2} = 36.45\,mol\,dm^{-3}$$

(c) The value of K_c will not change.

It is easy to be trapped into redoing the calculation, changing the concentrations to half their previous values and obtaining a new value for K_c. However, changing the volume of the container results in a change in the number of moles of each of the components so that the value of K_c remains the same. Remember, only a change in temperature causes a change in the value of K_c.

3 (a) If the pH of the acid is 2.70, $[H^+] = 10^{-2.7} = 2.0 \times 10^{-3}\,mol\,dm^{-3}$ (calculator value = $1.995262315 \times 10^{-3}$)

In the solution of ethanoic acid, $[H^+(aq)] = [CH_3COO^-(aq)]$

$$K_a = \frac{[H^+(aq)]^2}{[CH_3COOH(aq)]}$$

$$\text{Hence, } [CH_3COOH(aq)] = \frac{[H^+(aq)]^2}{K_a} = \frac{(2.0 \times 10^{-3})^2}{1.7 \times 10^{-5}} = 0.23(52) = 0.24\,mol\,dm^{-3}$$

If $[H^+(aq)]$ is not rounded in the first step, the concentration is $0.23\,mol\,dm^{-3}$. It is good practice to keep the values in the calculator and round off at the end of the calculation.

(b) For a buffer at pH 4.0, the concentration of $[H^+(aq)]$ is $1.0 \times 10^{-4}\,mol\,dm^{-3}$.

The concentration of the ethanoic acid stays the same, so use the expression:

$$K_a = \frac{[H^+(aq)][CH_3COO^-(aq)]}{[CH_3COOH(aq)]}$$

Rearranging gives:

$$\frac{K_a[CH_3COOH]}{[H^+]} = [CH_3COO^-]$$

$$= \frac{1.7 \times 10^{-5} \times 0.235}{1.0 \times 10^{-4}} = [CH_3COO^-] = 0.040\,mol\,dm^{-3}$$

Molar mass of sodium ethanoate = $82.0\,g\,mol^{-1}$

Therefore, mass required = $82.0 \times 0.040 = 3.28 = 3.3\,g$

4 (a) Indicators, HIn, are weak acids:

$HIn \rightleftharpoons H^+ + In^-$

$$K_{in} = \frac{[H^+][In^-]}{[HIn]}$$

At the mid-point, $[In^-] = [HIn]$

Hence, $K_{in} = [H^+]$

Therefore, $pK_{in} = pH$

$pH = -\log(6.31 \times 10^{-7}) = 6.20$

(b) $HIn \rightleftharpoons H^+ + In^-$

Yellow Red

In acid solution, the added H^+ pushes this equilibrium to the left and therefore the indicator changes colour to yellow.

In alkaline solution, the added OH^- reacts with the H^+ from the indicator to produce water. This causes the indicator to adjust by creating more In^- and, therefore, it changes colour to red.

At the end point, the colour of the indicator is orange.

(c) (i) $0.0001\,mol\,dm^{-3}$ hydrochloric acid has a pH of 4, so the indicator appears yellow.

(ii) Pure water has a pH of 7, so the indicator will be turning red. However, to the eye it will probably appear orange because conversion to the HIn form will be incomplete.

5 (a) At pH = 12, $[H^+(aq)] = 10^{-12}$ and $[OH^-(aq)] = \frac{K_w}{[H^+(aq)]} = 0.01\,mol\,dm^{-3}$

At pH = 11.5, $[H^+(aq)] = 10^{-11.5} = 3.16 \times 10^{-12}\,mol\,dm^{-3}$

$$\text{Therefore, } [OH^-(aq)] = \frac{10^{-14}}{3.16 \times 10^{-12}} = 0.00316 = 0.003\,mol\,dm^{-3}$$

Therefore, the change in concentration is $0.007\,mol\,dm^{-3}$.

(b) The indicator equilibrium is $NPB \rightleftharpoons H^+ + NPB^-$

At pH = 11.5, $[NPB] = [NPB^-]$

Therefore, $K_{in} = [H^+] = 10^{-11.5} = 3.16 \times 10^{-12} = 3 \times 10^{-12}\,mol\,dm^{-3}$

(c) $K_{in} = \frac{[H^+][NPB^-]}{[NPB]} = 3.16 \times 10^{-12}\ \text{mol dm}^{-3}$

At pH = 10.8, $[H^+] = 10^{-10.8}$

Therefore, $\frac{[10^{-10.8}][NPB^-]}{[NPB]} = 3.16 \times 10^{-12}$

$$\frac{[NPB^-]}{[NPB]} = \frac{3.16 \times 10^{-12}}{10^{-10.8}} = 0.20$$

Hence the ratio $[NPB]:[NPB^-]$ is 5.0:1.

6 (a) If the acid equilibrium for aspirin is written as: $HA(aq) \rightleftharpoons H^+(aq) + A^-(aq)$

$$\frac{[H^+(aq)][A^-(aq)]}{[HA(aq)]} = 3.0 \times 10^{-4}\ \text{mol dm}^{-3}$$

pH of the stomach is equal to 1, so $[H^+(aq)] = 0.1\ \text{mol dm}^{-3}$

ratio of the dissociated salt of aspirin : undissociated aspirin is:

$$\frac{[A^-(aq)]}{[HA(aq)]} = \frac{3.0 \times 10^{-4}}{0.1} = 3 \times 10^{-3}$$

This means the aspirin is largely in its molecular (undissociated) form, HA. This would be likely to dissolve in the lipids on the stomach lining, so bleeding might be a problem.

(b) At pH 7.4, $[H^+(aq)] = 4 \times 10^{-8}\ \text{mol dm}^{-3}$, $K_a = 3 \times 10^{-4}\ \text{mol dm}^{-3}$

So, $\frac{[A^-(aq)]}{[HA(aq)]} = \frac{3 \times 10^{-4}}{4 \times 10^{-8}} = 7.5 \times 10^3$

The aspirin is now largely ionised and is mostly present as the anion.

(c) The relative formula mass of $Ca(OH)_2$ is 74.1. Hence, the concentration of $Ca(OH)_2$ is 0.0100 mol dm^{-3}.

Calcium hydroxide is a strong base, so it will be ionised fully. The concentration of hydroxide ions present is 0.0200 mol dm^{-3}.

If $[OH^-(aq)] = 0.0200$, then $[H^+(aq)] = \frac{10^{-14}}{0.0200} = 5.0 \times 10^{-13}$

pH = 12.30

At this pH, there would be substantial base hydrolysis of the aspirin.

(d) $C_6H_4(OCOCH_3)CO_2H + H_2O \rightleftharpoons C_6H_4(OH)CO_2H + CH_3COOH$

At equilibrium, 0.117 g of CH_3COOH is present

moles of $CH_3COOH = \frac{0.117}{60.0} = 1.95 \times 10^{-3}$ mol

amount, in moles, of aspirin hydrolysed = 1.95×10^{-3} mol

initial amount, in moles, of aspirin = $\frac{0.900}{180.0} = 5.0 \times 10^{-3}$ mol

percentage of aspirin hydrolysed = $\frac{1.95 \times 10^{-3}}{5.00 \times 10^{-3}} \times 100 = 39.0\%$

5 Energy

Now test yourself

1 (a) ΔS will be negative (movement of particles in ice is more restricted)

(b) ΔS will be positive (the particles in an aqueous solution have more freedom than those in a solid)

(c) ΔS will be negative (the oxygen atoms from O_2 lose their freedom to move)

(d) ΔS will be negative (the reduction in overall volume as the reaction takes place reduces freedom of movement)

2 Substances with particles with the least freedom of movement tend to have the lowest entropy values. Of the three substances here, iodine is a solid and has particles with the least freedom to move. This identifies iodine as substance B with the entropy value 58.4 J mol^{-1} K^{-1}.

Methanol is a liquid and will have the next highest entropy value. Methanol is substance C with the entropy value 127.2 J mol^{-1} K^{-1}.

Ammonia is a gas and has particles with the greatest freedom to move. Ammonia is substance A with the entropy value 192.5 J mol^{-1} K^{-1}.

3 The equation for the reaction is:

$2Na(s) + \frac{1}{2}O_2(g) \rightarrow Na_2O(s)$

The entropy change, ΔS, is:

$72.8 - (2 \times 51.0 + \frac{1}{2} \times 102.5) = -80.5$ J mol^{-1} K^{-1}

4 (a) $CO_3^{2-}(aq) + 2H^+(aq) \rightarrow H_2O(l) + CO_2(g)$

(b) $CaCO_3(s) + 2H^+(aq) \rightarrow Ca^{2+}(aq) + H_2O(l) + CO_2(g)$

(c) $Ca^{2+}(aq) + CO_3^{2-}(aq) \rightarrow CaCO_3(s)$

(d) $OH^-(aq) + H^+(aq) \rightarrow H_2O(l)$

(e) $Cu^{2+}(aq) + 2OH^-(aq) \rightarrow Cu(OH)_2(s)$

(f) $ZnO(s) + 2H^+(aq) \rightarrow Zn^{2+}(aq) + H_2O(l)$

5 (a) Balance symbols and charge:

$MnO_4^-(aq) + 8H^+(aq) + 5e^- \rightarrow Mn^{2+}(aq) + 4H_2O(l)$

Balance charge:

$V^{2+}(aq) \rightarrow V^{3+}(aq) + e^-$

Multiply the second equation by 5 and add the equations together so that the electrons cancel:

$MnO_4^-(aq) + 8H^+(aq) + 5V^{2+}(aq) \rightarrow 5V^{3+}(aq) + Mn^{2+}(aq) + 4H_2O(l)$

(b) Balance symbols and charge:

$MnO_4^-(aq) + 8H^+(aq) + 5e^- \rightarrow Mn^{2+}(aq) + 4H_2O(l)$

Balance symbols and charge:

$V^{2+}(aq) + 3H_2O(l) \rightarrow VO_3^-(aq) + 6H^+(aq) + 3e^-$

Multiply the first equation by 3 and the second equation by 5, so that each has 15 electrons. Add together to give:

$3MnO_4^-(aq) + 24H^+(aq) + 5V^{2+}(aq) + 15H_2O(l) \rightarrow 5VO_3^-(aq) + 30H^+(aq) + 3Mn^{2+}(aq) + 12H_2O(l)$

Simplify by cancelling the H^+ and H_2O that appear on both sides of the equation:

$3MnO_4^-(aq) + 5V^{2+}(aq) + 3H_2O(l) \rightarrow 5VO_3^-(aq) + 6H^+(aq) + 3Mn^{2+}(aq)$

(c) Balance symbols and charge:

$Cr_2O_7^{2-}(aq) + 14H^+(aq) + 6e^- \rightarrow 2Cr^{3+}(aq) + 7H_2O(l)$

Balance symbols and charge:

$SO_2(aq) + 2H_2O(l) \rightarrow SO_4^{2-}(aq) + 4H^+(aq) + 2e^-$

Multiply the second equation by 3, then add together and simplify:

$Cr_2O_7^{2-}(aq) + 14H^+(aq) + 3SO_2(aq) + 6H_2O(l) \rightarrow 3SO_4^{2-}(aq) + 12H^+(aq) + 2Cr^{3+}(aq) + 7H_2O(l)$

This can be further simplified to:

$Cr_2O_7^{2-}(aq) + 2H^+(aq) + 3SO_2(aq) \rightarrow 3SO_4^{2-}(aq) + 2Cr^{3+}(aq) + H_2O(l)$

(d) Balance symbols and charge:

$NO_3^-(aq) + 4H^+(aq) + 3e^- \rightarrow NO(g) + 2H_2O(l)$

Balance charge:

$Cu(s) \rightarrow Cu^{2+}(aq) + 2e^-$

Multiply the first equation by 2 and the second equation by 3, so that each has six electrons. Add together to give:

$8H^+(aq) + 2NO_3^-(aq) + 3Cu(s) \rightarrow 3Cu^{2+}(aq) + 2NO(g) + 4H_2O(l)$

6 (a) $Mg(s) \rightarrow Mg^{2+}(aq) + 2e^-$ $E^\ominus = +2.37\,V$

$Zn^{2+}(aq) + 2e^- \rightarrow Zn(s)$ $E^\ominus = -0.76\,V$

Therefore, the overall cell potential is 1.61 V.

(b) $Fe^{3+}(aq) + e^- \rightarrow Fe^{2+}(aq)$ $E^\ominus = +0.77\,V$

$Sn^{2+}(aq) \rightarrow Sn^{4+}(aq) + 2e^-$ $E^\ominus = -0.15\,V$

Therefore, the overall cell potential is 0.62 V.

(c) $Br_2(aq) + 2e^- \rightarrow 2Br^-(aq)$ $E^\ominus = +1.09\,V$

$2I^-(aq) \rightarrow I_2(aq) + 2e^- E^\ominus = -0.54\,V$

Therefore, the overall cell potential is 0.55 V.

(d) $Zn(s) \rightarrow Zn^{2+}(aq) + 2e-$ $E^\ominus = +\,0.76\,V$

$I_2(aq) + 2e^- \rightarrow 2I^-(aq)$ $E^\ominus = +0.54\,V$

Therefore, the overall cell potential is 1.30 V.

(e) $Br_2(aq) + 2e^- \rightarrow 2Br^-(aq)$ $E^\ominus = +1.09\,V$

$Sn^{2+}(aq) \rightarrow Sn^{4+}(aq) + 2e^-$ $E^\ominus = -0.15\,V$

Therefore, the overall cell potential is 0.94 V.

Check your understanding

1 (a)

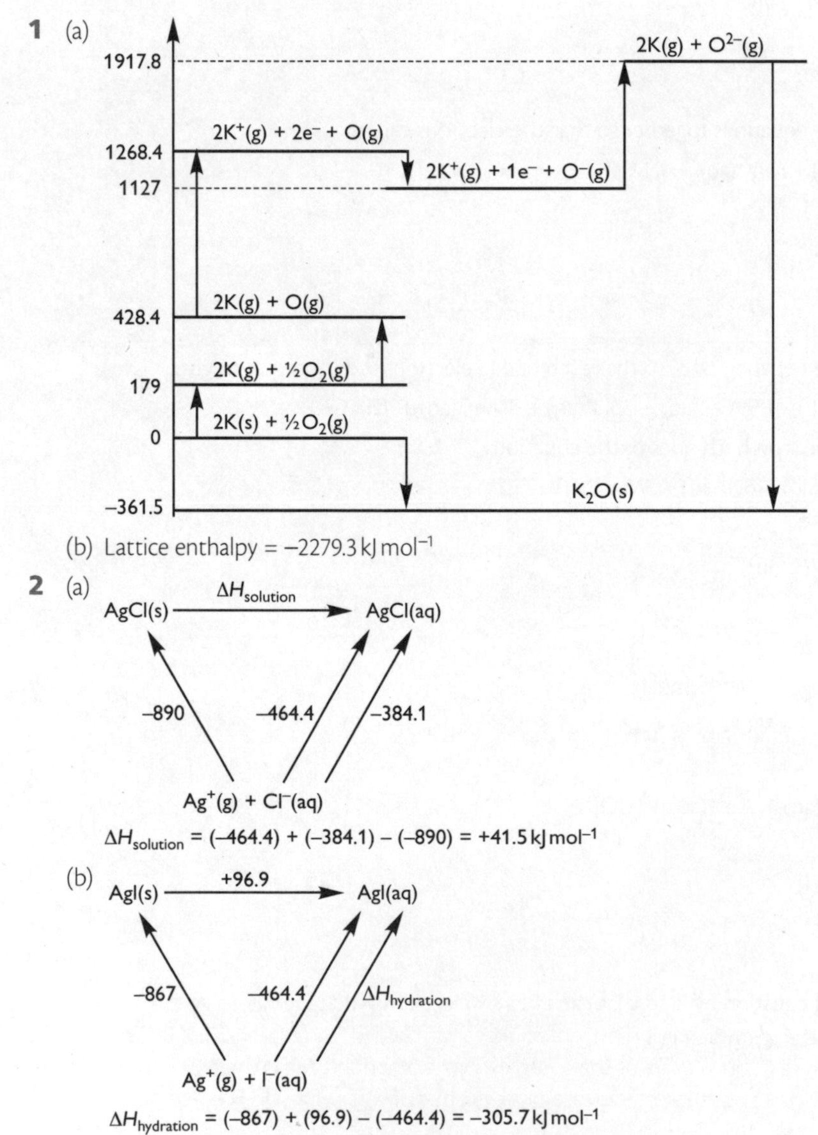

(b) Lattice enthalpy = $-2279.3\,kJ\,mol^{-1}$

2 (a)

$\Delta H_{solution} = (-464.4) + (-384.1) - (-890) = +41.5\,kJ\,mol^{-1}$

(b)

$\Delta H_{hydration} = (-867) + (96.9) - (-464.4) = -305.7\,kJ\,mol^{-1}$

(c) Enthalpies of hydration of Cl^- and I^- are = $-384.1\,kJ\,mol^{-1}$ and $-305.7\,kJ\,mol^{-1}$ respectively. The enthalpy of hydration of I^- is less negative than the enthalpy of hydration of Cl^- because the ionic radius of I^- is greater than that of Cl^-. This means that the attraction between I^- and the dipoles in H_2O is less than the attraction between Cl^- and the dipoles in H_2O.

(d) ΔH solution for both AgCl(aq) and AgI(aq) are endothermic. This indicates that neither substance will be soluble in water at standard temperature. ΔH solution for AgI is more endothermic than that for AgCl, indicating that AgI is less soluble than AgCl.

3 The aluminum foil sets up an electrical cell with the amalgam:

$Al(s) \rightarrow Al^{3+}(aq) + 3e^-$ +1.66 V

With saliva as the electrolyte, either of the two components of the amalgam is able to receive electrons. The cell potentials under standard conditions are as follows:

$Al(s) \rightarrow Al^{3+}(aq) + 3e^-$ +1.66 V

$Hg^+ + e^- \rightarrow Ag/Hg$ (amalgam)   +0.85 V

This has an overall potential of 2.51 V.

$Al(s) \rightarrow Al^{3+}(aq) + 3e^-$ +1.66 V

$Sn^{2+} + 2e^- \rightarrow Sn/Hg$ (amalgam)   −0.13 V

This has an overall potential of 1.53 V.

Although the conditions in the mouth are, of course, not standard conditions these electric cells are enough to cause a current to flow, resulting in a nasty shock.

6 Transition elements

Now test yourself

1 (a) The change in colour suggests that a ligand exchange has taken place. The aqueous copper sulfate solution is light blue because of the presence of the $[Cu(H_2O)_6]^{2+}$ ion. 1,2 diaminoethane is a stronger ligand than water and the change in colour to a darker blue is due to the formation of the copper-1,2 diaminoethane complex ion.

(b) When HCl is added, the hydrogen ions form a stronger bond with 1,2 diaminoethane than does the Cu^{2+} ion.

$H_2NCH_2CH_2NH_2 + 2H^+ \rightarrow {}^+H_3NCH_2CH_2NH_3{}^+$

The copper(II) 1,2 diaminoethane complex ion is broken down and the 1,2 diaminoethane is replaced by water molecules, restoring the original light blue colour.

2 (a) The aqueous cobalt ion is pink due to the presence of $[Co(H_2O)_6]^{2+}$ ions. However, at high concentration, chloride ions replace the water molecules, forming the blue cobalt chloride complex ion, $[CoCl_4]^{2-}$.

(b) In the presence of larger amounts of water, the process in part (a) is reversed and pink $[Co(H_2O)_6]^{2+}$ ions are reformed.

The equilibrium:

$[Co(H_2O)_6]^{2+} + 4Cl^- \rightleftharpoons [CoCl_4]^{2-} + 6H_2O$

moves readily from side-to-side depending on the concentration of chloride ions present (water is in vast excess).

(c) When aqueous silver nitrate is added, the silver ions react with chloride ions to produce a precipitate of silver chloride. The blue cobalt chloride complex ion, $[CoCl_4]^{2-}$ is destroyed and the pink $[Co(H_2O)_6]^{2+}$ ions are formed once again.

The Ag^+ ion removes the Cl^- ion from the equilibrium:

$[Co(H_2O)_6]^{2+} + 4Cl^- \rightleftharpoons [CoCl_4]^{2-} + 6H_2O$

Pink   Blue

The equilibrium moves to the left, hence the solution turns pink. AgCl(s) is also formed.

3 (a) oxidation number of Zn: +2

(b) oxidation number of Fe: +2

(c) oxidation number of Co: +3

(d) oxidation number of Co: +2

(e) oxidation number of Cr: +3

Check your understanding

1 (a) Compound **A** $[Co(NH_3)_6]Cl_3$ can release three Cl^- ions which will form a precipitate with added $Ag^+(aq)$ ions. Salt **B** $[Co(NH_3)_5Cl]Cl_2$ can only release two Cl^- ions because the third Cl^- ion is part of the complex ion . Compound **A** has three Cl^-; compound **B** has 2 Cl^-. Therefore compound **A** produces 3/2 times as much precipitate of silver chloride as compound **B** does.

Answers

(b) The compounds **C** and **D** must have only one Cl^- available as an anion and the other two Cl^- must be present in a complex cation with the four ammonia ligands. This suggests that **C** and **D** are *cis–trans* isomers.

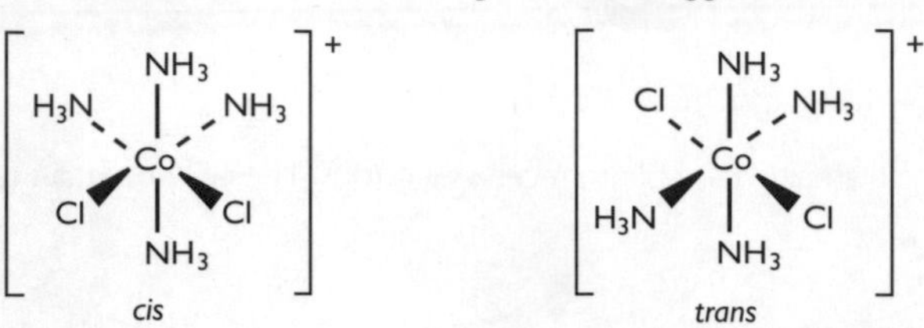

2 The half-equation for the reduction of $MnO_4^-(aq)$ is:

$MnO_4^-(aq) + 8H^+(aq) + 5e^- \rightarrow Mn^{2+}(aq) + 4H_2O(l)$

The half-equation for the oxidation of $VO^{2+}(aq)$ is:

$VO^{2+}(aq) + 2H_2O(l) \rightarrow VO_3^-(aq) + 4H^+(aq) + e^-$

Multiplying the second half-equation by 5, the overall equation for the reaction simplifies to:

$MnO_4^-(aq) + 5VO^{2+}(aq) + 6H_2O(l) \rightarrow 5VO_3^-(aq) + 12H^+(aq) + Mn^{2+}$

amount, in moles, of MnO_4^- in $23.3\,cm^3 = \frac{23.30}{1000} \times 0.0150 = 3.495 \times 10^{-4}\,mol$

Using the equation, the amount, in moles, of VO^{2+} used in the titration $= 5 \times 3.495 \times 10^{-4} = 1.7475 \times 10^{-3}\,mol$

Concentration of $VO^{2+} = \frac{1000}{25.0} \times 1.7475 \times 10^{-3} = 0.0699\,mol\,dm^{-3}$

3 (a) $Cr_2O_7^{2-}(aq) + 14H^+(aq) + 6e^- \rightarrow 2Cr^{3+}(aq) + 7H_2O(l)$

$Sn^{2+}(aq) \rightarrow Sn^{4+}(aq) + 2e^-$

The overall equation is obtained by multiplying the second half-equation by 3:

$Cr_2O_7^{2-}(aq) + 14H^+(aq) + 3Sn^{2+}(aq) \rightarrow 3Sn^{4+}(aq) + 2Cr^{3+}(aq) + 7H_2O(l)$

(b) amount, in moles, of $Cr_2O_7^{2-}$ in $20.0\,cm^3 = \frac{20.0}{1000} \times 0.0175 = 3.50 \times 10^{-4}\,mol$

Using the equation, the amount, in moles, of Sn^{2+} used in the titration $= 3 \times 3.50 \times 10^{-4} = 1.05 \times 10^{-3}\,mol$

concentration of $Sn^{2+}(aq) = \frac{1000}{25.0} \times 1.05 \times 10^{-3} = 0.0420\,mol\,dm^{-3}$

(c) The relative atomic mass of tin is 118.7.

mass of 0.0420 mol of tin $= 0.0420 \times 118.7 = 4.985\,g$

percentage by mass of tin in the solder $= \frac{4.985}{10.00} \times 100 = 49.9\%$

4 $(COOH)_2$ forms $(COO)_2^{2-}$ ions which are oxidised to CO_2

Hence:

$(COO)_2^{2-}(aq) \rightarrow 2CO_2(g) + 2e^-$

$MnO_4^-(aq) + 8H^+(aq) + 5e^- \rightarrow Mn^{2+}(aq) + 4H_2O(l)$

Multiplying the first half-equation by 5 and the second half-equation by 2 gives the overall equation:

$2MnO_4^-(aq) + 16H^+(aq) + 5(COO)_2^{2-}(aq) \rightarrow 10CO_2(g) + 2Mn^{2+}(aq) + 8H_2O(l)$

moles of MnO_4^- in $23.90\,cm^3 = \frac{23.90}{1000} \times 0.0200 = 4.78 \times 10^{-4}\,mol$

moles of $(COO)_2^{2-}$ used in the titration $= \frac{5}{2} \times 4.78 \times 10^{-4} = 1.195 \times 10^{-3}\,mol$

moles of $(COO)_2^{2-}$ in $250\,cm^3 = \frac{250}{25.0} \times 1.195 \times 10^{-3} = 0.0120\,mol$

The relative molecular mass of ethanedioic acid is 90.0.

mass of ethanedioic acid in four rhubarb leaves $= 90.0 \times 0.0120 = 1.08\,g$

$\therefore$ mass of ethanedioic acid in one rhubarb leaf $= 1.08/4 = 0.27\,g$

If 24 g is a fatal dose, the number of rhubarb leaves needed $= \frac{24}{0.27} = 89$ leaves.

5 Oxidation state of X in $NaXO_3$ is +5.

$25.0\,cm^3$ of $NaXO_3$ contains $\frac{25.0}{1000} \times 0.0500 = 1.25 \times 10^{-3}$ mol

The overall equation for the titration is:

$2S_2O_3^{2-}(aq) + I_2(aq) \rightarrow S_4O_6^{2-}(aq) + 2I^-(aq)$

relative formula mass of sodium thiosulfate = 158.2

concentration of sodium thiosulfate = $\frac{13.63}{158.2} = 0.08616\,mol\,dm^{-3}$

amount, in moles, of $S_2O_3^{2-}(aq)$ in $29.0\,cm^3 = \frac{29.00}{1000} \times 0.08616 = 2.50 \times 10^{-3}$ mol

amount, in moles, of iodine produced = 1.25×10^{-3} mol

so, in the reaction, 1 mol of XO_3^- produces 1 mol of I_2

Since $2I^-(aq) \rightarrow I_2(aq) + 2e^-$, the oxidation number of X in XO_3^- must reduce by 2 from +5 to +3.